Edexcel AS/A level

BIOLOGY B

1

nn Fullick

VAYS LEARNING

PEARSON

Published by Pearson Education Limited, 80 Strand, London WC2R 0RL.

www.pearsonschoolsandfecolleges.co.uk

Copies of official specifications for all Edexcel qualifications may be found on the website: www.edexcel.com

Text © Ann Fullick
Exam-style questions © Pearson Education Limited
Edited by Natalie Bayne and Jo Egré
Designed by Elizabeth Arnoux for Pearson Education Limited
Typeset by Tech-Set Ltd, Gateshead
Original illustrations © Pearson Education Limited 2015
Illustrated by Tech-Set Ltd, Gateshead and Peter Bull Art Studio
Cover design by Elizabeth Arnoux for Pearson Education Limited
Picture research by Caitlin Swain
Cover photo/illustration © Science Photo Library/King's College London

The rights of Ann Fullick and Graham Hartland to be identified as authors of this work have been asserted by them in accordance with the Copyright, Designs and Patents Act 1988.

First published 2008
Second edition published 2015

19 18 17 16
10 9 8 7 6 5 4

British Library Cataloguing in Publication Data
A catalogue record for this book is available from the British Library

ISBN 9781447991144

Printed in Italy by Lego SpA

Acknowledgements
Every effort has been made to contact copyright holders of material reproduced in this book. Any omissions will be rectified in subsequent printings if notice is given to the publishers.

A note from the publisher
In order to ensure that this resource offers high-quality support for the associated Edexcel qualification, it has been through a review process by the awarding body to confirm that it fully covers the teaching and learning content of the specification or part of a specification at which it is aimed, and demonstrates an appropriate balance between the development of subject skills, knowledge and understanding, in addition to preparation for assessment.

While the publishers have made every attempt to ensure that advice on the qualification and its assessment is accurate, the official specification and associated assessment guidance materials are the only authoritative source of information and should always be referred to for definitive guidance.

Edexcel examiners have not contributed to any sections in this resource relevant to examination papers for which they have responsibility.

No material from an endorsed book will be used verbatim in any assessment set by Edexcel.

Endorsement of a book does not mean that the book is required to achieve this Edexcel qualification, nor does it mean that it is the only suitable material available to support the qualification, and any resource lists produced by the awarding body shall include this and other appropriate resources.

Picture credits

The publisher would like to thank the following for their kind permission to reproduce their photographs:

(Key: b-bottom; c-centre; l-left; r-right; t-top)

Alamy Images: Ann and Steve Toon 168cr, Arco Images GmbH 208, BSIP SA 73br, Custom Life Science Images 33, Edwin Remsberg 70br, INTERFOTO 70bl, Jeremy Sutton-Hibbert 102, Mark Conlin 239, Nigel Cattlin 243bl, Patrick J. Endres 150–151, Picture Partners 120cl, Roberto Nistri 186, Scott Camazine 248, Steve Bloom Images 198, Stone Nature Photography 184c, The Natural History Museum 160br, 168cl; **Anthony Short:** 8–9, 10cr, 13, 26br, 110–111, 118, 126–127, 152, 153, 155tl, 155cl, 156t, 159tl, 159bl, 163b, 172–173, 175, 176bl, 179, 184cl, 188, 192–193, 194l, 196b, 204tr, 206cl, 213, 224–225, 232, 233, 238br, 290; **Ardea:** Bill Coster 178br/b, John Mason 178br/t; **Biodiversity Institute of Ontario:** 159r; **Copperhead Institute:** Chuck Smith 119; **Corbis:** Carolina Biological/Visuals Unlimited 116, CDC/PHIL 160cr, Viaframe 36–37; **David Grémillet:** 177tl, 177bl; **DK Images:** Lucy Claxton 156bl; **FLPA Images of Nature:** Fritz Polking 206cr; **Fotolia. com:** Simone Werner-Ney 202, tomatito26 26bl; **Getty Images:** Chris Jackson 205, De Agostini Picture Library 288, Dr. Brad Mogen 292, Glyn Kirk/AFP 204br, Matt Cardy 203, moodboard 133, Peter Tsai 54–55, Photolibrary 76cl, Photolibrary/Ed Reschke 83tl, Ralph Slepecky/Visuals Unlimited, Inc. 95tr, Roland Birke 226, Vetta 29; **Pearson Education Ltd:** 223tl; **Photoshot Holdings Limited:** Oceans-Image 251; **Phototake, Inc:** ISM 83bl; **Professor Legesse Negash:** BBC 207; **Science Photo Library Ltd:** A. Dowsett, Health Protection Agency 106, 70bc, 74, 130/2, 244–245, 258, 264cr, Adrian Bicker 156br, AMI Images 138, Andrew Lambert Photography 20, 31, Asa Thoresen 88bl, Athenais, ISM 264tl, Biology Pics 85, Biophoto Associates 84, 88br, 90bl, 130 (all), 176r, 255br, Chuck Brown 254, CNRI 90tl, 99, D. Phillips 140, 255tr, David M. Phillips 217c, David McCarthy 122, David Scharf 117, Dirk Wiersma 157, Don W. Fawcett 78tl, Dr. P. Marazzi 261, Dr. Yorgos Nikas 143, Dr. Gopal Murti 76bl, 82, 90r, Dr. Jeremy Burgess 86, 87, 237, Dr. John Brackenbury 176cl, Dr. Kari Lounatmaa 75tr, Dr. Keith Wheeler 238bl, Dr. Richard Kessel & Dr. Gene Shih, Visuals Unlimited 243br, Dr. Rosalind King 95br, Dr. Stanley Flegler/Visuals Unlimited, Inc. 217cl, 217bl, Dr. Stanley Flegler, Visuals Unlimited 210–211, Eye of Science 16–17, 101, 182cr, Herve Conge, ISM 124bl, 278, Innerspace Imaging 88bc, J.C. Revy, ISM 217tr, Jackie Lewin, Royal Free Hospital 51, James King-Holmes 146, Juan Gaertner 40, Lee D. Simon 100, Look at Sciences 112, Louise Hughes 56, Louise Murray 155tr, M.I. Walker 121, Martin Oeggerli 92–93, Martin Shields 22, Martyn F. Chillmaid 18, Medimage 23b, Mehau Kulyk 120tl, Michael Abbey 217cr, NIBSC 75bl, Omikron 98, Pascal Goetgheluck 162, PHOTOTAKE Inc. 78tr, Power and Syred 23t, 91, 279, Pr. G. Gimenez-Martin 124tl, Professor P. Motta & D. Palermo 115, Professors P. Motta & T. Naguro 77, Richard J. Green 95cr, Science Source 221, Scott Camazine 285, Sovereign, ISM 265, Steve Gschmeissner 72, 73cr, 231, 253, Ted Kinsman 243bc, Thomas Ames Jr., Visuals Unlimited 243tr, Thomas Deerinck, NCMIR 68–69, Tom Kinsbergen 182tl; **Shutterstock. com:** Alexey Repka 154, Debra James 206tr, Egon Zitter 194c, idreamphoto 12, Image Point Fr 267, Jeff Dalton 155cr, Jim Lopes 284, Joel Blit 163t, Nicky Rhodes 197, Picsfive 272, Svitlana S. 144, Vlad61 194r; **The University of California:** Alex McPherson, Irvine/National Institute of General Medical Sciences 57; **U.S. Department of Agriculture:** Agricultural Research Service 158; **Veer/Corbis:** Backyard Productions 174, enjoylife25 183, gbrouwer 200, goce risteski 128, luchschen 276–277, marilyna 64, Nyker 184cr, prochasson frederic 10bl; **Wellcome Trust Sanger Institute:** 41

Cover images: *Front:* **Science Photo Library Ltd:** King's College London

All other images © Pearson Education Limited

Picture Research by: Caitlin Swain

We are grateful to the following for permission to reproduce copyright material:

Figures

Figure on page 32 from 'Trehalose: an intriguing disaccharide with potential for medical application in ophthalmology', Clinical ophthalmology, 5, 577 (2011), Clinical Ophthalmology by Society for Clinical Ophthalmology (Great Britain) Reproduced with permission of Dove Medical Press Limited in the format Republish in a book via Copyright Clearance Center; Figure on page 203 from the front cover of the DEFRA publication 'What nature can do for you', https://www.gov.uk/government/uploads/system/uploads/attachment_data/file/221097/pb13897-nature-do-for-you.pdf, Published by the Department for Environment, Food and Rural Affairs. © Crown Copyright 2010; Figure on page 240 from http://www.abpischools.org.uk/page/modules/breathingandasthma/asthma7.cfm, ABPI Resources for Schools, Association of the British Pharmaceutical Industry (ABPI) with permission.

Text

Article on page 32 from 'Trehalose: an intriguing disaccharide with potential for medical application in ophthalmology', Clinical ophthalmology, 5, 577 (2011), Clinical Ophthalmology by Society for Clinical Ophthalmology (Great Britain) Reproduced with permission of Dove Medical Press Limited in the format Republish in a book via Copyright Clearance Center; Article on page 106 adapted from 'Deadly Ebola virus "could spread globally" after plane brings it to Nigeria', Daily Mail 28/07/2014 (Nick Fagge), Daily Mail; Article on page 106 adapted from 'Epidemiology and surveillance', http://www.afro.who.int/en/clusters-a-programmes/dpc/epidemic-a-pandemic-alert-and-response/outbreak-news/4236-ebola-virus-disease-west-africa-29-july-2014.html, © Copyright World Health Organization (WHO) – Regional Office for Africa, 2013. All rights reserved. Article on page 106 from http://www.wales.nhs.uk/sitesplus/888/page/74608, Public Health Wales; Extract on page 146 adapted from 'In vitro fertilisation', Heinemann Library (Fullick, A.); Poetry on page 168 from 'Oxford Ragwort' (Short, G.), with permission from Anthony Short; Article on page 188 adapted from 'Quagga rebreeding: a success story', Farmer's Weekly (Harvey, K.), © 2014 Farmer's Weekly Magazine; Article on page 188 from 'A rapid loss of stripes: the evolutionary history of the extinct quagga', September 2005 Volume: 1 Issue: 3 (Jennifer A. Leonard et al), Copyright © 2014, The Royal Society; Article on page 240 from http://www.abpischools.org.uk/page/modules/breathingandasthma/asthma7.cfm, ABPI Resources for Schools, Association of the British Pharmaceutical Industry (ABPI) with permission; Article on page 290 from Encyclopedia of Life Sciences, John Wiley & Sons, Ltd (Turgor Pressure by Jeremy Pritchard, University of Birmingham 2001) © 2001, John Wiley & Sons, Ltd, Reproduced with permission of Blackwell Publishing.

The Publisher would like to thank Chris Curtis and Wade Nottingham for their contributions to the Maths skills section of this book.

The author would like to acknowledge and thank the teams at Science and Plants for Schools (SAPS), the Wellcome Trust Sanger Institute and the ABPI for their valuable input. The author would also like to thank the following for their support and individual contributions: Dr Jeremy Pritchard; Alice Kelly; Amy Ekins-Coward; Tony Short; William Fullick; Thomas Fullick; James Fullick, Edward Fullick; Chris Short.

Every effort has been made to contact copyright holders of material reproduced in this book. Any omissions will be rectified in subsequent printings if notice is given to the publishers.

Contents

How to use this book

Welcome to your Edexcel AS/A level Biology B course. In this book you will find a number of features designed to support your learning.

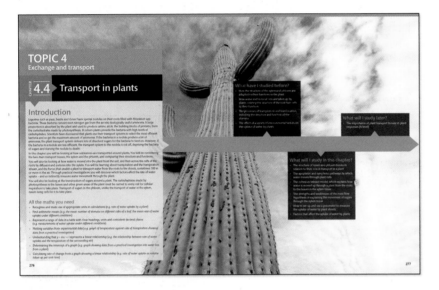

Chapter openers

Each chapter starts by setting the context for that chapter's learning:

- Links to other areas of Biology are shown, including previous knowledge that is built on in the chapter, and future learning that you will cover later in your course.

- The **All the maths you need** checklist helps you to know what maths skills will be required.

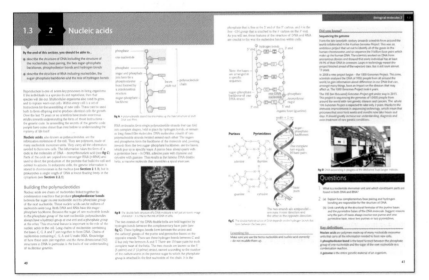

Main content

The main part of each chapter covers all the points from the specification that you need to learn. The text is supported by diagrams and photos that will help you understand the concepts.

Within each section, you will find the following features:

- **Learning objectives** at the beginning of each section, highlighting what you need to know and understand.

- **Key definitions** shown in bold and collated at the end of each section for easy reference.

- **Worked examples** showing you how to work through questions, and how your calculations should be set out.

- **Learning tips** to help you focus your learning and avoid common errors.

- **Did you know?** boxes featuring interesting facts to help you remember the key concepts.

- **Questions** to help you check whether you have understood what you have just read, and whether there is anything that you need to look at again.

Thinking Bigger

The book features a number of **Thinking Bigger** spreads that give you an opportunity to read and work with real-life research and writing about science. The timeline at the bottom of the spreads highlights which of the chapters the material relates to. These spreads will help you to:

- read real-life material that's relevant to your course
- analyse how scientists write
- think critically and consider the issues
- develop your own writing
- understand how different aspects of your learning piece together.

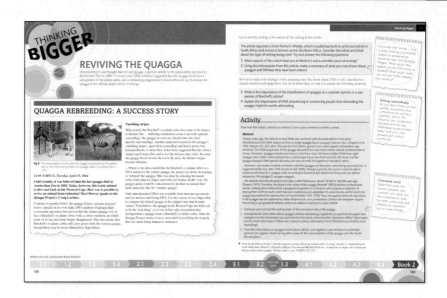

Exam-style questions

At the end of each chapter there are also **exam-style questions** to help you to:

- test how fully you have understood the learning
- practise for your exams.

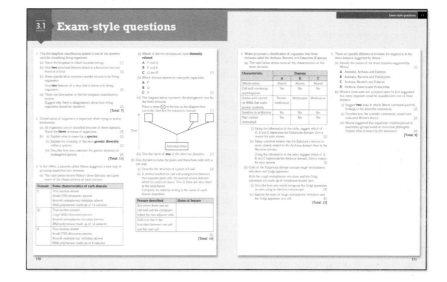

Getting the most from your online ActiveBook

This book comes with 3 years' access to ActiveBook* – an online, digital version of your textbook. Follow the instructions printed on the inside front cover to start using your ActiveBook.

Your ActiveBook is the perfect way to personalise your learning as you progress through your Edexcel AS/A level Biology course. You can:

- access your content online, anytime, anywhere
- use the inbuilt highlighting and annotation tools to personalise the content and make it really relevant to you
- search the content quickly using the index.

Highlight tool
Use this to pick out key terms or topics so you are ready and prepared for revision.

Annotations tool
Use this to add your own notes, for example links to your wider reading, such as websites or other files. Or make a note to remind yourself about work that you need to do.

*For new purchases only. If this access code has already been revealed, it may no longer be valid. If you have bought this textbook secondhand, the code may already have been used by the first owner of the book.

TOPIC 1
Biological molecules

1.1 > Chemistry for life

Introduction

A raft spider *Dolomedes fimbriatus* sits on the surface of the water, hidden by the stems of water plants, waiting for the vibrations in the surface tension that alert her to the presence of her prey. She is large – up to 23 mm across – yet water-repellent hairs enable her to run across the surface to grab her victims. These are usually aquatic invertebrates that also live on or near the water surface. Water is vital for this semiaquatic spider – and for all life on Earth.

Biology is the study of living things. The basic unit of life is the cell, and underpinning all life is – chemistry! The way atoms are bonded together affects the way chemicals work in the cells – and that affects everything, from the way plants make food by photosynthesis to the way your eyes respond to light.

In this chapter you will be looking at some of the key ways in which atoms and molecules interact to make up the chemistry of life. You will be using these basic principles throughout your biology course, because they underpin the structures and functions of all the organisms you will study.

Around two-thirds of the surface of the Earth is covered in water and around two-thirds of your body is water. The oceans, rivers and lakes of the world are teeming with life and all the reactions in your cells take place in solution in water. In this chapter you will be applying your knowledge of the basic chemical principles to help you understand just why water is so vital for life.

All the maths you need

- Recognise and make appropriate use of units in calculations (*e.g. millimetres*)
- Use ratios (*e.g. representing the relationships between atoms in an ion or molecule*)

What have I studied before?

- Life processes depend on molecules whose structure is related to their function
- All living things are made up of cells
- Many processes in living cells, including diffusion and osmosis, depend on water
- Reactions in cells take place in solution in water
- Water is needed for photosynthesis
- Water is transported around plants
- Water makes up around two-thirds of the human body

What will I study later?

- How carbohydrates, lipids and proteins are formed by covalent bonding
- The importance of polarity in the structure and function of phospholipids
- The importance of hydrogen bonding in the tertiary and quaternary structure of proteins and in the structure and function of enzymes
- How water is taken into and moved around plants
- The movement of water into and out of cells, tissues and vessels in animals, plants and fungi
- The importance of water in plant movements
- The role of water in the reactions of cellular respiration and photosynthesis (A level)

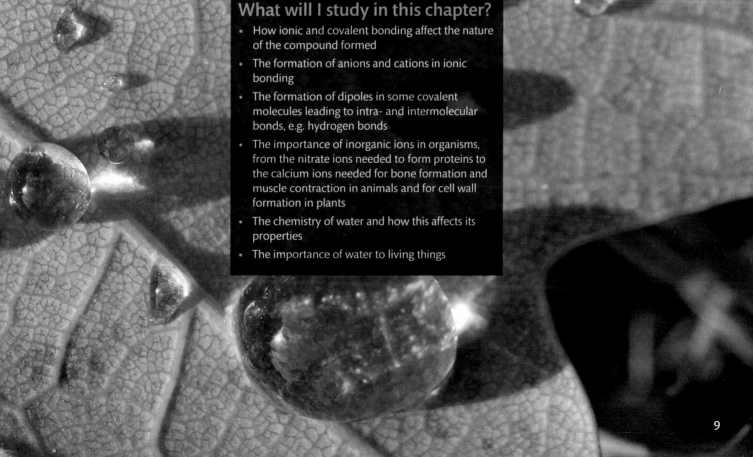

What will I study in this chapter?

- How ionic and covalent bonding affect the nature of the compound formed
- The formation of anions and cations in ionic bonding
- The formation of dipoles in some covalent molecules leading to intra- and intermolecular bonds, e.g. hydrogen bonds
- The importance of inorganic ions in organisms, from the nitrate ions needed to form proteins to the calcium ions needed for bone formation and muscle contraction in animals and for cell wall formation in plants
- The chemistry of water and how this affects its properties
- The importance of water to living things

By the end of this section, you should be able to...

● explain the role of inorganic ions in plants

● explain the importance of the dipole nature of water in the formation of hydrogen bonds and the significance of some of the properties of water to the organisms

Ionic and covalent bonding

Biology is the study of living things – but living things are made up of chemicals. If you understand some of the basic principles of chemistry, you will also develop a much better understanding of biological systems. The chemical bonds within and between molecules affect the properties of the compounds they form. This in turn affects their functions within the cell and the organism.

fig A All life depends on some very fundamental chemistry.

The single basic unit of all elements is the atom. When the atoms of two or more elements react they form a compound. An atom is made up of a nucleus containing positive protons and neutral neutrons surrounded by negative electrons. We model these electrons as orbiting around the nucleus in shells. When an atom has a full outer shell of electrons it is stable and does not react. However, most atoms do not have a full outer shell of electrons. In chemical reactions, they are involved in changes that give them a stable outer shell. There are two ways they can achieve this:

• **Ionic bonding:** the atoms involved in the reaction donate or receive electrons. The atom, or part of the molecule, gains one or more electrons and becomes a negative ion (**anion**). The other atom, or part of the molecule, loses one or more electrons and becomes a positive ion (**cation**). Strong forces of attraction called ionic bonds hold the oppositely charged ions together.

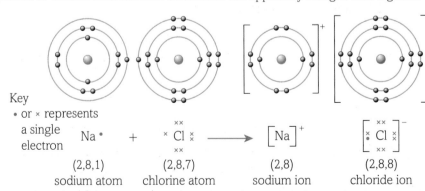

Key
• or × represents a single electron

Na • + × Cl × ⟶ [Na]⁺   [× Cl ×]⁻

(2,8,1) sodium atom   (2,8,7) chlorine atom   (2,8) sodium ion   (2,8,8) chloride ion

fig C Animals such as this cow can use a mineral lick to get the salt they need to function.

fig B The formation of sodium chloride (salt), an inorganic substance that is very important in living organisms, is an example of ionic bonding.

- **Covalent bonding:** the atoms involved in the reaction share electrons. Covalent bonds are very strong and the molecules formed are usually neutral. However, in some covalent compounds, the molecules are slightly polarised. The electrons in the covalent bonds are not quite evenly shared. This means the molecule has a part that is slightly negative and a part that is slightly positive. This separation of charge is called a **dipole**, and the tiny charges are represented as δ^+ and δ^- (see **fig D**). The molecule is described as a **polar molecule**. This polarity is particularly common if one or more hydrogen atoms are involved in the bond.

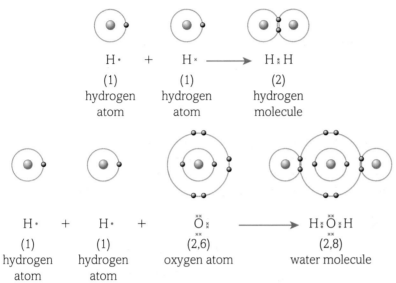

fig D The formation of hydrogen molecules and water molecules are examples of covalent bonding.

The importance of inorganic ions

When ionic substances are dissolved in water, the ions separate. Cells are 60–70% water, and so in living organisms most ionic substances exist as positive and negative ions. Many of these ions play specialised roles in individual cells and in the functioning of entire organisms. Here are some of the inorganic ions you will meet as you study biology, with an indication of one or more of their roles:

Important anions

- nitrate ions (NO_3^-) – needed in plants for the formation of amino acids and therefore proteins from the products of photosynthesis, and also for the formation of DNA

- phosphate ions (PO_4^{3-}) – needed in all living organisms including plants and animals in the formation of ATP and ADP as well as DNA and RNA

- chloride ions (Cl^-) – needed in nerve impulses and many secretory systems

- hydrogen carbonate ions (HCO_3^-) – needed for buffering the blood to prevent it from becoming too acidic

Important cations

- sodium ions (Na^+) – needed in nerve impulses and many secretory systems

- calcium ions (Ca^{2+}) – needed for the formation of calcium pectate for the middle lamella between two cell walls in plants, and for bone formation and muscle contraction in animals

- hydrogen ions (H^+) – needed in cellular respiration and photosynthesis, and in numerous pumps and systems in organisms as well as pH balance

- magnesium ions (Mg^{2+}) – needed for production of chlorophyll in plants

The chemistry of water

Water is the medium in which all the reactions take place in living cells. Without it, substances could not move around the body. Water is one of the reactants in the process of photosynthesis, on which almost all life depends. And water is a major habitat – it supports more life than any other part of the planet. Understanding the properties of water will help you understand many key systems in living organisms.

fig E Water is vital for life on Earth in many different ways.

The importance of water to biological systems is due to the basic chemistry of its molecules. The simple chemical formula of water is H_2O. This tells us that two atoms of hydrogen are joined to one atom of oxygen to make up each water molecule (see **fig F**). However, because the electrons are held closer to the oxygen atom than to the hydrogen atoms, water is a polar molecule.

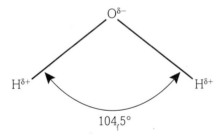

fig F A model of a water molecule.

One of the most important results of this charge separation is that water molecules form **hydrogen bonds**. The slightly negative oxygen atom of one water molecule will attract the slightly positive hydrogen atoms of other water molecules in a weak electrostatic attraction called a hydrogen bond. This means that the molecules of water 'stick together' more than you might otherwise expect, because although each individual hydrogen bond is weak, there are a great many of them (as shown in **fig G**). Water has relatively high melting and boiling points compared with other substances that have molecules of a similar size – it takes more energy to overcome the attractive forces of all the hydrogen bonds. Hydrogen bonds are an important concept in biochemistry – for example they play an important part in protein structure (see **Section 1.2.4**) and in the structure and functioning of DNA (see **Section 1.3.2**).

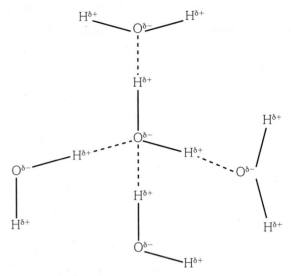

fig G Hydrogen bonding in water molecules.

The importance of water

The properties of water make it very important in biological systems for several reasons:

- Water is a polar solvent. Because water is a polar molecule many ionic substances like sodium chloride will dissolve in it. Many covalently bonded substances are also polar and they too will dissolve in water, although they often do not dissolve in other covalently bonded solvents such as ethanol. As a result most of the chemical reactions within cells occur in water (in aqueous solution).

- Water is an excellent transport medium because so many different substances will dissolve in it. Water also carries other substances such as starch that form colloids rather than solutions.

- As water cools to 4 °C, it reaches its maximum density. As it cools further, the molecules become more widely spaced. As a result, ice is less dense than water and floats, forming an insulating layer and helping to prevent the water underneath it from freezing. It also melts quickly because it is at the top, exposed to the sun. It is very unusual for the solid form of a chemical to be less dense than the liquid, but as a result of this unusual property, organisms can live in water even in countries where it gets cold enough to freeze in winter.

- Water is slow to absorb and release heat – it has a high specific heat capacity. The hydrogen bonds between the molecules mean it takes a lot of energy to separate them. This means the temperature of large bodies of water such as lakes and seas does not change much throughout the year, making them good habitats for living organisms.

- Water is a liquid and so it cannot be compressed. This is an important factor in many hydraulic mechanisms in living organisms.

- Water molecules are cohesive – the forces between the molecules mean they stick together. This is very important for the movement of water from the roots to the leaves of plants.

• Water molecules are adhesive – they are attracted to other different molecules. This is also important in plant transport systems and in surface tension.

• Water has a very high surface tension because the attraction between the water molecules, including hydrogen bonds, is greater than the attraction between the water molecules and the air. As a result the water molecules hold together forming a thin skin of surface tension. Surface tension is of great importance in plant transport systems, and also affects life at the surface of ponds, lakes and other water masses.

fig H Without surface tension a raft spider could not move across the water in this way.

Questions

1 How do ionic bonds and covalent bonds differ?

2 What are the differences between ionic substances and polar substances?

3 How are hydrogen bonds formed between water molecules and what effect do they have on the properties of water?

4 The properties of water affect its role in living organisms. Discuss.

Key definitions

An **anion** is a negative ion, formed when an atom gains electron(s).

A **cation** is a positive ion, formed when an atom loses electron(s).

Ionic bonds are attractive forces between oppositely charged ions.

Covalent bonds are formed when atoms share electrons.

A **dipole** is the separation of charge in a molecule when the electrons in covalent bonds are not evenly shared.

A **polar molecule** is a molecule containing a dipole.

Hydrogen bonds are weak electrostatic intermolecular bonds formed between polar molecules containing at least one hydrogen atom.

Biology has a lot of application of scientific knowledge, so it's a good idea to remind yourself of the basics learnt at GCSE.

1 Remind yourself of ionic bonds by answering these questions.

(a) Draw a diagram of a sodium atom, including the protons, neutrons and electrons. [2]

(b) Draw a diagram of a chlorine atom, including the protons, neutrons and electrons. [2]

(c) Now show how sodium and chlorine atoms can be turned into sodium and chloride ions to form the ionic bond. [2]

[Total: 6]

2 (a) Draw one water molecule. [1]

(b) Using the atomic structure of oxygen and hydrogen, explain why the electrons are held closer to the oxygen atom. [2]

(c) Explain how a molecule of sodium chloride can dissolve in water. [3]

[Total: 6]

3 Read through the following account about water, then write on the dotted lines the most appropriate word or words to complete the account.

Water molecules are described as .... because they have a slight positive charge at one end of the molecule and a slight negative charge at the other end. This makes water a good .... for salts and substances such as sugars.

Bonds that form between water molecules are called .... bonds.

Water is a good coolant because it has a high ...., which means that it takes a lot of heat to change it from a liquid to a gas.

Water also has a high ...., which means that a lot of energy is needed to cause a small rise in its temperature. [5]

[Total: 5]

4 Fill in this table to show which ion is used for which purpose.

Name of ion	Symbol	Function in plants
Nitrate		
	PO^{2-}	
	Ca^{2+}	
		Needed to produce chlorophyll

[4]

[Total: 4]

5 There are many substances important to living organisms. These can be classified as

A cations
B anions
C polar molecules
D non-polar molecules

Identify the following molecules using one of the terms above.

(a) water

(b) chloride ion (Cl^-)

(c) sodium ion (Na^+)

(d) hydrogen carbonate ion

(e) methane

(f) phosphate ion [6]

[Total: 6]

6 Acids release hydrogen ions (H^+) into solution. Explain how hydrogen carbonate ions (HCO_3^-) act to prevent the blood becoming too acidic. [2]

[Total: 2]

7 (a) A student wrote a title to her table of results in a water-based ink, and then underlined in ballpoint pen. Her lab partner then accidentally spilled water over the page. The title smudged, but the underlining didn't. Using your knowledge of the properties of water, explain these observations. [2]

(b) Having done some research, the student decided that it would be more sensible to do her tables of results using a pencil. Use your knowledge of solvents to explain why this is a good idea. [2]

[Total: 4]

8 (a) Draw the electron shells of the following atoms:
        (i) carbon
        (ii) oxygen
        (iii) sodium
        (iv) argon   [4]
    (b) Use the information from the electron shells to state how
        many protons each of the above elements has.   [4]
    (c) Use the information of the number of protons to explain
        why CH_4 is a non-polar molecule but H_2O is a polar
        molecule.   [2]
    (d) Use the periodic table to find the relative atomic mass of
        each element. Why is this number always bigger than the
        proton number?   [1]
    (e) Looking again at the electron shells, explain why carbon
        can form four bonds, oxygen can form two, sodium only
        forms one bond, but argon can form no bonds.   [4]
                                                            [Total: 15]

9   Marion wanted to build a pond to breed fish in the north of
    England. Temperatures in this region can fall below 0 °C in
    the winter. She was advised to make sure the pond was at
    least 3 m deep and held 3 500 000 litres of water. Use your
    knowledge of the properties of water to explain why such a
    large pond was necessary.   [4]
                                                            [Total: 4]

10 Pond skaters are insects that can travel on the surface of water.
    Using your knowledge of the properties of water, explain how
    these insects can travel like this.   [3]
                                                            [Total: 3]

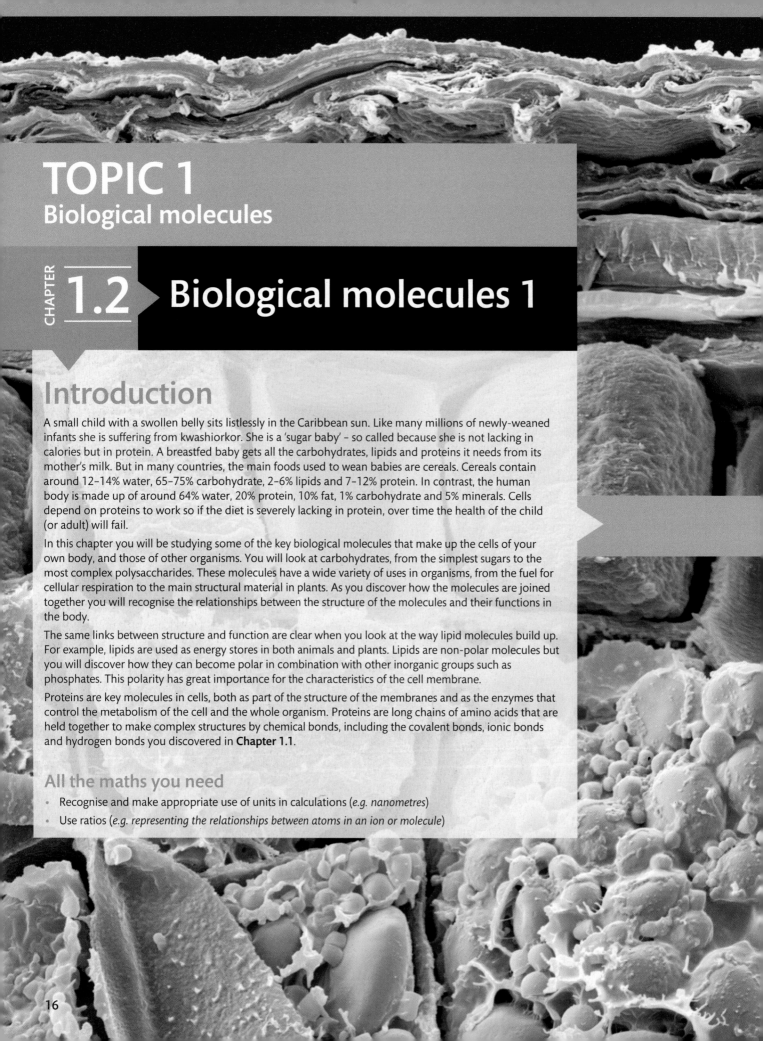

TOPIC 1
Biological molecules

Introduction

A small child with a swollen belly sits listlessly in the Caribbean sun. Like many millions of newly-weaned infants she is suffering from kwashiorkor. She is a 'sugar baby' – so called because she is not lacking in calories but in protein. A breastfed baby gets all the carbohydrates, lipids and proteins it needs from its mother's milk. But in many countries, the main foods used to wean babies are cereals. Cereals contain around 12–14% water, 65–75% carbohydrate, 2–6% lipids and 7–12% protein. In contrast, the human body is made up of around 64% water, 20% protein, 10% fat, 1% carbohydrate and 5% minerals. Cells depend on proteins to work so if the diet is severely lacking in protein, over time the health of the child (or adult) will fail.

In this chapter you will be studying some of the key biological molecules that make up the cells of your own body, and those of other organisms. You will look at carbohydrates, from the simplest sugars to the most complex polysaccharides. These molecules have a wide variety of uses in organisms, from the fuel for cellular respiration to the main structural material in plants. As you discover how the molecules are joined together you will recognise the relationships between the structure of the molecules and their functions in the body.

The same links between structure and function are clear when you look at the way lipid molecules build up. For example, lipids are used as energy stores in both animals and plants. Lipids are non-polar molecules but you will discover how they can become polar in combination with other inorganic groups such as phosphates. This polarity has great importance for the characteristics of the cell membrane.

Proteins are key molecules in cells, both as part of the structure of the membranes and as the enzymes that control the metabolism of the cell and the whole organism. Proteins are long chains of amino acids that are held together to make complex structures by chemical bonds, including the covalent bonds, ionic bonds and hydrogen bonds you discovered in **Chapter 1.1**.

All the maths you need

- Recognise and make appropriate use of units in calculations (*e.g. nanometres*)
- Use ratios (*e.g. representing the relationships between atoms in an ion or molecule*)

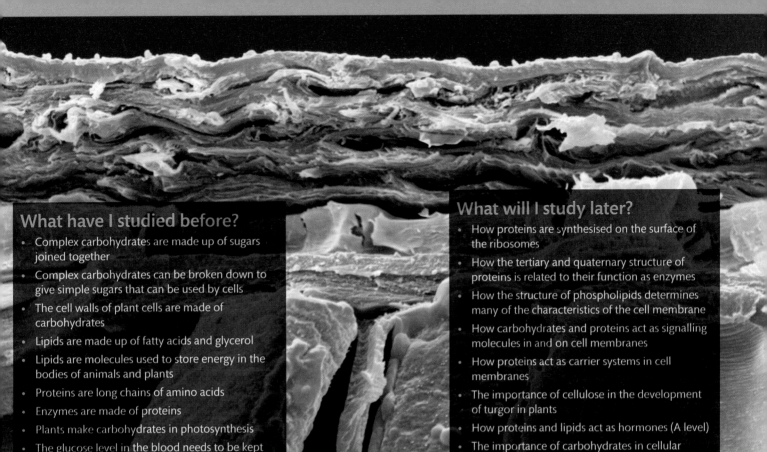

What have I studied before?

- Complex carbohydrates are made up of sugars joined together
- Complex carbohydrates can be broken down to give simple sugars that can be used by cells
- The cell walls of plant cells are made of carbohydrates
- Lipids are made up of fatty acids and glycerol
- Lipids are molecules used to store energy in the bodies of animals and plants
- Proteins are long chains of amino acids
- Enzymes are made of proteins
- Plants make carbohydrates in photosynthesis
- The glucose level in the blood needs to be kept within tight limits

What will I study later?

- How proteins are synthesised on the surface of the ribosomes
- How the tertiary and quaternary structure of proteins is related to their function as enzymes
- How the structure of phospholipids determines many of the characteristics of the cell membrane
- How carbohydrates and proteins act as signalling molecules in and on cell membranes
- How proteins act as carrier systems in cell membranes
- The importance of cellulose in the development of turgor in plants
- How proteins and lipids act as hormones (A level)
- The importance of carbohydrates in cellular respiration (A level)

What will I study in this chapter?

- What makes an organic compound
- The structure of different types of monosaccharides
- The formation of disaccharides by the joining of two monosaccharides in a condensation reaction
- The structure of complex polysaccharides and how their structure is related to their functions as storage molecules and structural compounds
- The role of lipids in cell membranes and their importance as storage molecules in plants and animals
- The structure of lipids including the formation of ester bonds
- The structure of amino acids, peptides and polypeptides and how they relate to each other
- The formation of peptide bonds between amino acids
- The primary, secondary, tertiary and quaternary structure of proteins and how the structure is related to the function of the protein

Carbohydrates 1 – monosaccharides and disaccharides

By the end of this section, you should be able to...

● describe the difference between monosaccharides and disaccharides

● describe the structure of the hexose glucose (alpha and beta) and the pentose ribose

● explain how monosaccharides join to form disaccharides through condensation reactions forming glycosidic bonds, and how they can be split through hydrolysis reactions

● explain how the structure of glucose relates to its function

What are organic compounds?

Biological molecules are the key to the structure and function of living things. Biological molecules are often organic compounds. Organic compounds all contain carbon atoms. They also contain atoms of hydrogen, oxygen and, less frequently, nitrogen, sulfur and phosphorus. Most of the material in your body that is not water is made up of these organic molecules. An understanding of why organic molecules are special will help you to understand the chemistry of biological molecules including carbohydrates, lipids and proteins.

Each carbon atom can make four bonds and so it can join up with four other atoms. Carbon atoms bond particularly strongly to other carbon atoms to make long chains. The four bonds of a carbon atom usually form a tetrahedral shape and this leads to the formation of branched chains, or rings, or any number of three-dimensional (3D) shapes. In some carbon compounds small molecules (**monomers**) bond with many other similar units to make a very large molecule called a **polymer**. The ability of carbon to combine and make **macromolecules** (large molecules) is the basis of all biological molecules and provides the great variety and complexity found in living things.

These two bonds are in the plane of the paper.

This bond goes back behind the plane of the paper.

This bond sticks out of the plane of the paper.

This part of a chain molecule:

can be shown with corners representing carbons, and its hydrogens ignored:

or more often as:

fig A The bonds in a carbon atom have a complicated 3D shape. This is difficult to represent, so in most molecular diagrams we use one of several different ways to draw them 'flat'.

fig B Carbohydrates are important molecules in plants and animals alike – and they also play a major role in the human diet.

Carbohydrates

Carbohydrates are important in cells as a usable energy source. They are also used for storing energy, and in plants, fungi and bacteria they form an important part of the cell wall. The best known carbohydrates are sugars and **starch**. **Sucrose** is the white crystalline sugar familiar to us all, while **glucose** is the energy supplier in sports and health drinks. Starch is found in flour and potatoes. But the group of chemicals known as carbohydrates contains many more compounds, as you will discover.

The basic structure of all carbohydrates is the same. They are made up of carbon, hydrogen and oxygen. There are three main groups of carbohydrates with varying complexity of molecules: **monosaccharides**, **disaccharides** and **polysaccharides**.

Monosaccharides – the simple sugars

Monosaccharides are simple sugars in which there is one oxygen atom and two hydrogen atoms for each carbon atom present in the molecule. A general formula for this can be written $(CH_2O)_n$. Here n can be any number, but it is usually low:

- **Triose sugars** ($n=3$) have three carbon atoms and the molecular formula $C_3H_6O_3$. They are important in the mitochondria, where glucose is broken down into triose sugars during respiration.

- **Pentose sugars** ($n=5$) have five carbon atoms and the molecular formula $C_5H_{10}O_5$. **Ribose** and **deoxyribose** are important in the nucleic acids **deoxyribonucleic acid (DNA)** and **ribonucleic acid (RNA)**, which make up the genetic material (see **Sections 1.3.1 and 1.3.2**).

- **Hexose sugars** ($n=6$) have six carbon atoms and the molecular formula $C_6H_{12}O_6$. They are the best known monosaccharides, often taste sweet and include glucose, galactose and fructose.

Molecular formulae show you how many atoms there are in the molecule, and what type they are, but they do not tell you what the molecule looks like and why it behaves as it does. To show this you can use displayed formulae. Although these do not follow every wiggle and kink in the carbon chain, they can give you a good idea of how the molecules are arranged in three dimensions. This can reveal all sorts of secrets about why biological systems behave as they do (see **fig D**).

ribose

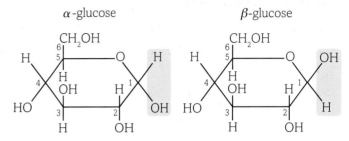

fig C Pentose sugars such as ribose have 5 carbon atoms.

α-glucose   **fructose**

fig D Hexose sugars have a ring structure. The arrangement of the atoms on the side chains can make a significant difference to the way in which the molecule can be used by the body. We number the carbon atoms so we can identify the different arrangements.

α-glucose and β-glucose

Glucose comes in different forms (**isomers**), including α-glucose and β-glucose. These two isomers result from different arrangements of the atoms on the side chains of the molecule (see **fig E**). The different isomers form different bonds between neighbouring glucose molecules, and this affects the polymers that are made.

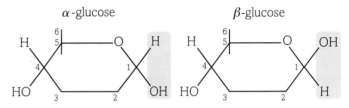

or, even more simply:

In these diagrams, the positions of carbon atoms are represented by their numbers only.

Note carefully the different arrangement of atoms around the carbon 1 atom in α-glucose and β-glucose.

fig E The difference in structure between α-glucose and β-glucose may seem small, but it has a big impact on the function of each molecule.

Did you know?

Hydrogenating some sugars reduces the energy they provide. When glucose is hydrogenated it forms sorbitol ($C_6H_{14}O_6$). Sorbitol tastes up to 60% sweeter than glucose but it provides less energy when it is used in the body (11 kJg^{-1} compared to 17 kJg^{-1}). The combination of the very sweet taste and the lower energy count makes it useful as a sweetener for people who want to lose weight. A small change in the chemical structure has a big effect on function.

Disaccharides – the double sugars

Disaccharides are made up of two monosaccharides joined together – for example sucrose (ordinary table sugar) is formed by a molecule of α-glucose joining with a molecule of fructose. Two monosaccharides join in a **condensation reaction** to form a disaccharide, and a molecule of water (H_2O) is removed. The link between the two monosaccharides results in a covalent bond known as a **glycosidic bond** (see **fig F**). We use numbers to show which carbon molecules are involved in the bond. If carbon 1 on one monosaccharide joins to carbon 4 on another monosaccharide, we call it a 1,4-glycosidic bond. If the bond is between carbon 1 and carbon 6, it is a 1,6-glycosidic bond.

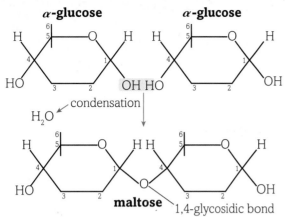

fig F The formation of a glycosidic bond. The condensation reaction between two monosaccharides results in a disaccharide and a molecule of water.

When different monosaccharides join together, different disaccharides result. Many disaccharides taste sweet.

Disaccharide	Source	Monosaccharide
sucrose	stored in plants such as sugar cane	α-glucose + fructose
lactose	milk sugar – this is the main carbohydrate found in milk	α-glucose + β-galactose
maltose	malt sugar – found in germinating seed such as barley	α-glucose + α-glucose

table A Three common disaccharides.

Did you know?

Testing for sugars

- Benedict's solution is a chemical test for **reducing sugars**. It is a bright blue solution that contains copper(II) ions. Some sugars react readily with this solution when heated gently and reduce the copper(II) ions to copper(I) ions, forming a precipitate and giving a colour change from blue to orange. They are known as reducing sugars. All of the monosaccharides and some disaccharides are reducing sugars.
- Some sugars do not react with Benedict's solution. They are known as **non-reducing sugars**. You can heat a non-reducing sugar such as sucrose with a few drops of hydrochloric acid to hydrolyse the glycosidic bonds. Allow it to cool and then neutralise the solution with sodium hydrogen carbonate. This produces the monosaccharide units of the sugar, which will now give a positive Benedict's test.

fig G Benedict's test for reducing sugars.

Questions

1 What are the properties of organic compounds that make them so important in living organisms?

2 Describe how a glycosidic bond is formed between two monosaccharides to form a disaccharide.

Key definitions

A **monomer** is a small molecule that is a single unit of a larger molecule called a polymer.

A **polymer** is a long chain molecule made up of many smaller, repeating monomer units joined together by chemical bonds.

A **macromolecule** is a very large molecule often formed by polymerisation.

Starch is a long chain polymer formed of α-glucose monomers.

Sucrose is a sweet tasting disaccharide formed by the joining of α-glucose and fructose by a glycosidic bond.

Glucose is a hexose sugar.

A **monosaccharide** is a single sugar monomer.

A **disaccharide** is a sugar made up of two monosaccharide units joined by a glycosidic bond, formed in a condensation reaction.

A **polysaccharide** is a polymer made up of long chains of monosaccharide units joined by glycosidic bonds.

A **triose sugar** is a sugar with three carbon atoms.

A **pentose sugar** is a sugar with five carbon atoms.

Ribose is a pentose sugar that makes up part of the structure of RNA.

Deoxyribose is a pentose sugar that makes up part of the structure of DNA.

Deoxyribonucleic acid (DNA) is a nucleic acid that acts as the genetic material in many organisms.

Ribonucleic acid (RNA) is a nucleic acid which can act as the genetic material in some organisms and is involved in protein synthesis.

A **hexose sugar** is a sugar with six carbon atoms.

Isomers are molecules that have the same chemical formula, but different molecular structures.

A **condensation reaction** is a reaction in which a molecule of water is removed from the reacting molecules as a bond is formed between them.

A **glycosidic bond** is a covalent bond formed between two monosaccharides in a condensation reaction.

Reducing sugars are sugars that react with blue Benedict's solution and reduce the copper(II) ions to copper(I) ions giving an orangey-red precipitate.

Non-reducing sugars are sugars that do not react with Benedict's solution.

Carbohydrates 2 – polysaccharides

By the end of this section, you should be able to...

● explain how monosaccharides join to form polysaccharides through condensation reactions forming glycosidic bonds; and how these can be split through hydrolysis reactions

● explain how the structure of polysaccharides relates to their functions

The most complex carbohydrates are the polysaccharides. They are made of many monosaccharide units joined by condensation reactions that form glycosidic bonds (see **Section 1.2.1, fig F**). Molecules with 3–10 sugar units are known as **oligosaccharides**, while molecules containing 11 or more monosaccharides are known as true polysaccharides. Polysaccharides do not have the sweet taste of many mono- and disaccharides, but these complex polymers form some very important biological molecules.

The structure of polysaccharides makes them ideal as storage molecules:

- They can form very compact molecules, so large numbers can be stored in a cell.

- The glycosidic bonds are easily broken, allowing rapid release of monosaccharide units for cellular respiration.

- They are not very soluble in water, so have little effect on water potential within a cell and cause no osmotic water movements.

The glycosidic bond between two monosaccharides is split by a process known as **hydrolysis** (see **fig A**). The hydrolysis reaction is the opposite of the condensation reaction that formed the molecule, so water is added to the bond. Polysaccharides are gradually broken down into shorter and shorter chains and eventually single sugars are left. Disaccharides break down to form two monosaccharides. Hydrolysis takes place during digestion in the gut, and also in the muscle and liver cells when the carbohydrate stores are broken down to release sugars for use in cellular respiration.

Learning tip

Remember that glycosidic bonds are formed with the removal of a molecule of water in condensation reactions and broken with the addition of a molecule of water in hydrolysis reactions.

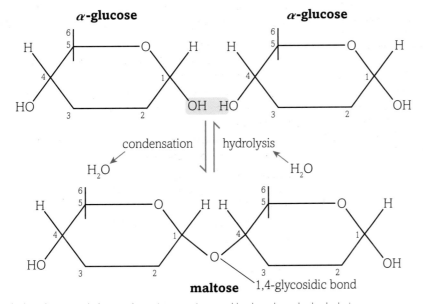

fig A Glycosidic bonds are made by condensation reactions and broken down by hydrolysis.

Carbohydrates as energy stores

Starch

Starch is particularly important as an energy store in plants. The sugars produced by photosynthesis are rapidly converted into starch, which is insoluble and compact but can be broken down rapidly to release glucose when it is needed. Storage organs such as potatoes are particularly rich in starch.

Starch is made up of long chains of α-glucose. But if you look at it more closely you will see that it is actually a mixture of two compounds:

Amylose: an unbranched polymer made up of between 200 and 5000 glucose molecules. As the chain lengthens the molecule spirals, which makes it more compact for storage.

Amylopectin: a branched polymer of glucose molecules. The branching chains have many terminal glucose molecules that can be broken off rapidly when energy is needed.

Amylose and amylopectin are both long chains of α-glucose molecules – so why are the molecules so different? It all depends on the carbon atoms involved in the glycosidic bonds.

Amylose is made up purely of α-glucose molecules joined by 1,4-glycosidic bonds, which is why the molecules are long unbranched chains.

In amylopectin many of the glucose molecules are joined by 1,4-glycosidic bonds, but there are also a few 1,6-glycosidic bonds. This results in the branching chains that change the properties of the molecule.

So starch has a combination of straight chain amylose and branched chain amylopectin molecules. This combination explains why carbohydrate foods like pasta are so good for you when you are doing sport. The amylopectin releases glucose for cellular respiration rapidly when needed. Amylose releases glucose more slowly over a longer period, keeping you going longer.

fig B Amylose and amylopectin – a small difference in the position of the glycosidic bonds in the molecule makes a big difference to the properties of the compounds.

Glycogen

Glycogen is sometimes referred to as 'animal starch' because it is the only carbohydrate energy store found in animals (see **fig C**). It is also an important storage carbohydrate in fungi. Chemically, glycogen is very similar to the amylopectin molecules in starch, and is also made up of many α-glucose units. Like starch, it is very compact, but the glycogen molecule has more 1,6-glycosidic bonds, giving it many side branches. As a result, glycogen can be broken down very rapidly. This makes it an ideal source of glucose for active tissues with a constantly high rate of cellular respiration, such as muscle and liver tissue.

Carbohydrates in plants

Polysaccharides are very important in plants. Starch is the main energy storage material in plants. A typical starch grain in a plant cell contains 70–80% amylopectin, with the rest being amylose.

(a) starch grains in a plant cell

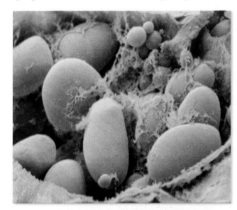

(b) glycogen granules in liver cells

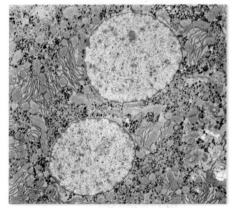

fig C Storage carbohydrates in plant and animal cells.

Cellulose is an important structural material in plants. The cell wall (see **Section 2.1.5**) is an important feature that gives plants their strength and support. It is made up largely of insoluble cellulose. Cellulose has much in common with starch and glycogen. It consists of long chains of glucose joined by glycosidic bonds. However, as you will remember, there are two structural isomers of glucose, α-glucose and β-glucose.

In starch, the monomer units are α-glucose. In cellulose, they are β-glucose and are held together by 1,4-glycosidic bonds where one of the monomer units has to be turned round (inverted) so the bonding can take place. This linking of β-glucose molecules means that the hydroxyl (–OH) groups stick out on both sides of the molecule (see **fig D**). This means hydrogen bonds can

form between the partially positively charged hydrogen atoms of the hydroxyl groups and the partially negatively charged oxygen atoms in other areas of the glucose molecules. This is known as cross-linking and it holds neighbouring chains firmly together.

Many of these hydrogen bonds form, making cellulose a material with considerable strength. Cellulose molecules do not coil or spiral – they remain as very long, straight chains. In contrast, starch molecules, with 1,4- and 1,6-glycosidic bonds between α-glucose monomers, form compact globular molecules that are useful for storage.

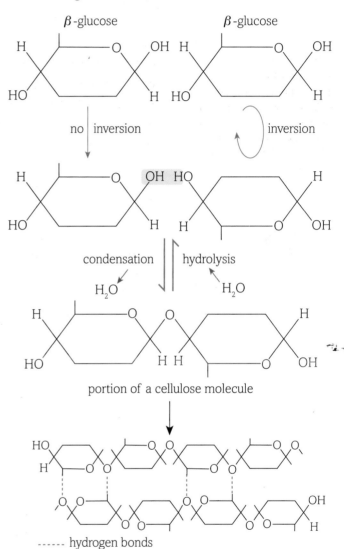

fig D Cellulose molecules consist of β-glucose monomers joined together by 1,4-glycosidic bonds.

This difference in structure between starch and cellulose gives them very different properties and functions. Starch is an important source of energy in the diet for many animals. However, most animals do not possess the enzymes needed to break the 1,4-glycosidic bonds between the molecules of β-glucose and so they cannot digest cellulose. Ruminants such as cows and sheep, have bacteria, fungi and protozoa living in their gut which produce cellulose-digesting enzymes. It is the cellulose in plant food that acts as roughage or fibre in the human diet – an important part of a healthy diet even though you cannot digest it.

Questions

1 Explain how the structure of carbohydrates is related to their function as storage molecules providing the fuel for cellular respiration in animals and plants.

2 Explain how the chemical structure of cellulose differs from that of starch and how this affects the way they can be used to supply energy in animals.

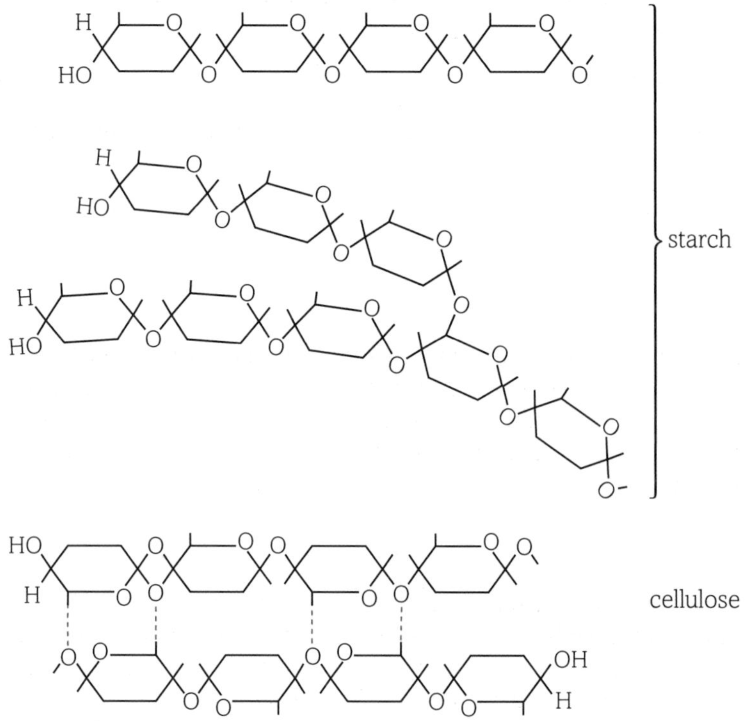

starch

cellulose

Key definitions

Oligosaccharides are molecules with 3–10 monosaccharide units.

Hydrolysis is a reaction in which bonds are broken by the addition of a molecule of water.

Amylose is a complex carbohydrate containing only glucose monomers joined together by 1,4-glycosidic bonds so the molecules form long unbranched chains.

Amylopectin is a complex carbohydrate made up of glucose monomers joined by both 1,4-glycosidic bonds and 1,6-glycosidic bonds so the molecules branch repeatedly.

Glycogen is made up of many α-glucose units joined by 1,4-glycosidic bonds but also has 1,6-glycosidic bonds, giving it many side branches.

Cellulose is a complex carbohydrate with β-glucose monomers held together by 1,4-glycosidic bonds. It is very important in plant cell walls.

Lipids

By the end of this section, you should be able to...

● explain the synthesis of a triglyceride, including the formation of ester bonds during condensation reactions between glycerol and three fatty acids

● describe the differences between unsaturated and saturated fatty acids

● explain how the structure of lipids relates to their role in energy storage, waterproofing and insulation

● explain the structure and properties of phospholipids in relation to their function in the cell membranes

The **lipids** are another group of organic chemicals that play a vital role in organisms. They form an integral part of all cell membranes and are also used as an energy store. Many plants and animals convert spare food into oils or fats to use when they are needed. For example, the seeds of plants contain lipids to provide energy for the seedling when it starts to grow, which is why seeds are such an important food source for many animals.

Fats and oils

Fats and oils are important groups of lipids. Chemically they are extremely similar, but fats, such as butter, are solids at room temperature and oils, such as olive oil, are liquids. Like carbohydrates, all lipid molecules contain the chemical elements carbon, hydrogen and oxygen. However, lipids contain a considerably lower proportion of oxygen than carbohydrates. Fats and oils are made up of two types of organic chemicals, **fatty acids** and **glycerol** (propane-1,2,3-triol). They are combined using **ester bonds**. Glycerol has the chemical formula $C_3H_8O_3$ (see **fig A**).

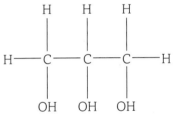

fig A Displayed formula of glycerol (propane-1,2,3-triol).

All fatty acids have a long hydrocarbon chain – a pleated backbone of carbon atoms with hydrogen atoms attached, and a carboxyl group (–COOH) at one end.

Living tissues contain more than 70 different kinds of fatty acids. Fatty acids vary in two ways:

• The length of the carbon chain can differ (although often 15–17 carbon atoms long in organisms).

• The fatty acid may be a **saturated fatty acid** or **unsaturated fatty acid**.

In a saturated fatty acid, each carbon atom is joined to the one next to it by a single covalent bond. A common example is stearic acid (see **fig B**). In an unsaturated fatty acid, the carbon chains have one or more double covalent bonds in them. A **monounsaturated fatty acid** has one double bond and a **polyunsaturated fatty acid** has more than one double bond (see **fig C**). Linoleic acid is an example of a polyunsaturated fatty acid. It is an essential fatty acid in our diet because we cannot make it from other chemicals.

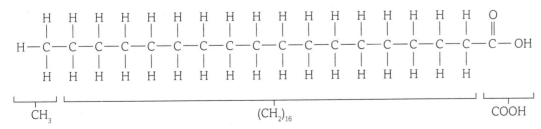

fig B Displayed formula of stearic acid, a saturated fatty acid found in both plant and animal fats.

$$H-\overset{\overset{\displaystyle H}{|}}{\underset{\underset{\displaystyle H}{|}}{C}}-\overset{\overset{\displaystyle H}{|}}{\underset{\underset{\displaystyle H}{|}}{C}}-\overset{\overset{\displaystyle H}{|}}{\underset{\underset{\displaystyle H}{|}}{C}}-\overset{\overset{\displaystyle H}{|}}{\underset{\underset{\displaystyle H}{|}}{C}}-\overset{\overset{\displaystyle H}{|}}{\underset{\underset{\displaystyle H}{|}}{C}}-\overset{\overset{\displaystyle H}{|}}{C}=\overset{\overset{\displaystyle H}{|}}{C}-\cdots$$

fig C Displayed formula of linoleic acid, a polyunsaturated fatty acid.

Forming ester bonds

A fat or oil results when glycerol combines with one, two or three fatty acids to form a monoglyceride, a diglyceride or a triglyceride. A bond is formed in a condensation reaction between the carboxyl group (–COOH) of a fatty acid and one of the hydroxyl groups (–OH) of the glycerol. A molecule of water is removed and the resulting bond is known as an ester bond. This type of condensation reaction is called **esterification** (see **fig D**). The nature of the lipid formed depends on which fatty acids are present. So, for example, lipids containing saturated fatty acids are more likely to be solid at room temperature than those containing unsaturated fatty acids.

For simplicity, fatty acids are represented by this general formula where 'R' represents the hydrocarbon chain. The fatty acids below are drawn in reversed form.

$$R-\overset{\overset{\displaystyle O}{\|}}{C}-OH$$

fig D Formation of ester bonds.

Note: there are only 6 atoms of oxygen in a triglyceride molecule.

The nature of lipids

Lipids contain many carbon-hydrogen bonds and little oxygen. When lipids are oxidised in respiration, the bonds are broken and carbon dioxide and water are the ultimate products. This reaction can be used to drive the production of a lot of ATP (see **Section 1.3.1**). Lipids, especially triglycerides, store about three times as much energy as the same mass of carbohydrates.

The **hydrophobic** nature of lipids is a key feature of their role in waterproofing organisms. Oils are important in waterproofing the fur and feathers of mammals and birds, while insects and plants use waxes for waterproofing their outer surfaces (see **fig E**). Lipids are good insulators – a fatty sheath insulates your nerves so the electrical impulses travel faster. They also insulate animals against heat loss – the thick layer of blubber in whales is a good example. Lipids have a very low density, so the body fat of water mammals helps them to float easily. All lipids dissolve in organic solvents, but are insoluble in water, so lipids do not interfere with the many water-based reactions that go on in the cytoplasm of a cell.

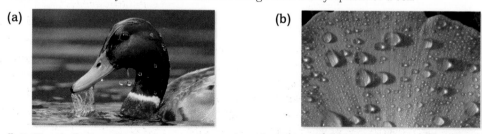

fig E Oil on the feathers of birds and the waxy layer on the surface of these ginkgo leaves makes them very waterproof.

Phospholipids

Inorganic phosphate ions ($-PO_4^{3-}$) are present in the cytoplasm of every cell. Sometimes one of the hydroxyl groups of glycerol undergoes an esterification reaction with a phosphate group instead of with a fatty acid, and a simple **phospholipid** is formed. Phospholipids are important because the lipid and the phosphate parts of the molecule give it very different properties.

The fatty acid chains of a phospholipid are neutral and insoluble in water. In contrast, the phosphate head carries a small negative charge and is soluble in water. When these phospholipids come into contact with water, the two parts of the molecule behave differently. The polar phosphate part is **hydrophilic** and dissolves readily in water (see **fig F**). The lipid tails are hydrophobic, so they do not dissolve in water. If the molecules are tightly packed in water they either form a **monolayer**, with the hydrophilic heads in the water and the hydrophobic lipid tails in the air, or clusters called **micelles**. In a micelle, all the hydrophilic heads point outwards and all the hydrophobic tails are inside (see **fig G**).

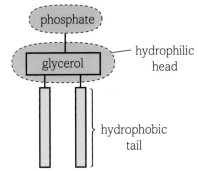

fig F A phospholipid.

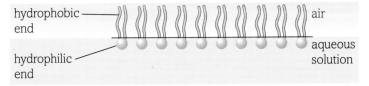

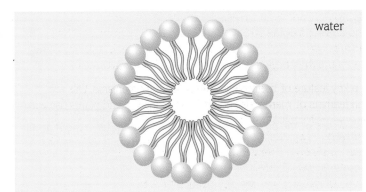

fig G Phospholipids form a monolayer at an air/water junction and a micelle when water surrounds them.

A phospholipid monolayer may form at a surface between air and water, but this is a fairly rare situation in living cells where there are water-based solutions on either side of the membranes. With water on each side, the phospholipid molecules form a **bilayer** with the hydrophilic heads pointing into the water, protecting the hydrophobic tails in the middle (see **fig H**). This structure, the **unit membrane**, is the basis of all membranes.

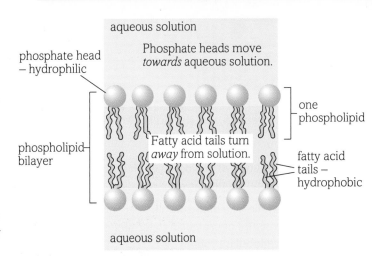

fig H A lipid bilayer is the backbone of all membrane structures in a cell.

Questions

1 Describe the main difference between a saturated and an unsaturated lipid, and the effect of this difference on the properties of the lipids.

2 Explain how triglycerides are formed.

Key definitions

Lipids are a large family of organic molecules that are important in cell membranes and as an energy store in many organisms. They include triglycerides, phospholipids and steroids.

A **fatty acid** is an organic acid with a long hydrocarbon chain.

Glycerol is propane-1,2,3-triol, an important component of triglycerides.

An **ester bond** is a bond formed in a condensation reaction between the carboxyl group (-COOH) of a fatty acid and one of the hydroxyl groups (-OH) of glycerol.

A **saturated fatty acid** is a fatty acid in which each carbon atom is joined to the one next to it in the hydrocarbon chain by a single covalent bond.

An **unsaturated fatty acid** is a fatty acid in which the carbon atoms in the hydrocarbon chain have one or more double covalent bonds in them.

A **monounsaturated fatty acid** is a fatty acid with only one double covalent bond between carbon atoms in the hydrocarbon chain.

A **polyunsaturated fatty acid** is a fatty acid with two or more double covalent bonds between carbon atoms in the hydrocarbon chain.

Esterification is the formation of ester bonds.

A **phospholipid** is a chemical in which glycerol bonds with two fatty acids and an inorganic phosphate group.

Hydrophilic molecules dissolve readily in water.

Hydrophobic molecules will not dissolve in water.

A **monolayer** is a single closely packed layer of atoms or molecules.

A **micelle** is a spherical aggregate of molecules in water with hydrophobic areas in the middle and hydrophilic areas outside.

A **bilayer** is a double layer of closely packed atoms or molecules.

A **unit membrane** is a bilayer structure formed by phospholipids in an aqueous environment, with the hydrophobic tails in the middle and the hydrophilic heads on the outside.

By the end of this section, you should be able to...

- outline the structure of an amino acid
- explain the formation of polypeptides and proteins and the nature of the bonds in proteins
- explain the significance of the primary, secondary, tertiary and quaternary structure of protein in determining the properties of fibrous and globular proteins
- explain how the structure of collagen and haemoglobin is related to their function

About 18% of your body is made up of protein. Proteins form hair, skin and nails, the enzymes needed for metabolism and digestion, and many of the hormones that control various body systems. They enable muscle fibres to contract, form antibodies that protect you from disease, help clot your blood and transport oxygen in the form of **haemoglobin**. Understanding the structure of proteins helps you develop an insight into the detailed biology of cells and organisms. Like carbohydrates and lipids, proteins contain carbon, hydrogen and oxygen. In addition they all contain nitrogen and many proteins also contain sulfur.

Proteins are another group of macromolecules made up of many small monomer units called **amino acids** joined together by condensation reactions. Amino acids combine in long chains to produce proteins. There are about 20 different naturally occurring amino acids that can combine in different ways to form a vast range of different proteins.

Amino acids

All amino acids have the same basic structure, which is represented as a general formula. There is always an amino group ($-NH_2$) and a carboxyl group ($-COOH$) attached to a carbon atom (see **fig A**). The group known as the R group varies between amino acids. This is where sulfur and selenium are found in the structure of a few amino acids. The structure of the R group affects the way the amino acid bonds with others in the protein, depending largely on whether the R group is polar or not.

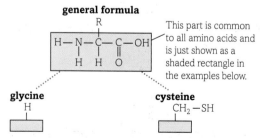

fig A Some different amino acids. In the simplest amino acid, glycine, R is a single hydrogen atom. In a larger amino acid such as cysteine, R is much more complex.

Forming proteins from amino acids

Amino acids join together by a reaction between the amino group of one amino acid, and the carboxyl group of another. They join in a condensation reaction and a molecule of water is lost. A **peptide bond** is formed when two amino acids join and a **dipeptide** is the result (see **fig B**). The R group is not involved in this reaction. More and more amino acids join to form **polypeptide** chains, which contain from a hundred to many thousands of amino acids. When the polypeptide folds or coils or associates with other polypeptide chains, it forms a protein.

amino acid 1　　　　　**amino acid 2** (inverted)

condensation　　hydrolysis

H_2O　　　　　H_2O

peptide bond

dipeptide

fig B Amino acids are the building blocks of proteins, joined together by peptide bonds.

Bonds in proteins

The peptide bond between amino acids is a strong bond. Other bonds also form between the amino acids in a chain to form the 3D structure of the protein. They depend on the atoms in the R group and include hydrogen bonds, **disulfide bonds** and ionic bonds.

Hydrogen bonds

In amino acids, tiny negative charges are present on the oxygen of the carboxyl groups and tiny positive charges are present on the hydrogen atoms of the amino groups. When these charged groups are close to each other, the opposite charges attract, forming a hydrogen bond. Hydrogen bonds are weak, but they can potentially form between any two amino acids positioned correctly, so there are lots of them holding the protein together very firmly. Hydrogen bonds break easily and reform if pH or temperature conditions change. They are very important in the folding and coiling of the polypeptide chains (see **fig C**).

Disulfide bonds

Disulfide bonds form when two cysteine molecules are close together in the structure of a polypeptide (see **fig C**). An oxidation reaction takes place between the two sulfur containing groups, resulting in a strong covalent bond known as a disulfide bond. These disulfide bonds are much stronger than hydrogen bonds, but they occur much less often. They are important for holding the folded polypeptide chains in place.

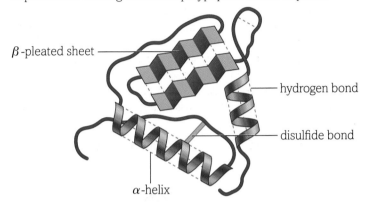

fig C Hydrogen bonds and disulfide bonds maintain the shape of protein molecules and this determines their function.

Ionic bonds

Ionic bonds can form between some of the strongly positive and negative amino acid side chains found buried deep in the protein molecules. These links are known as salt bridges. They are strong bonds, but they are not as common as the other structural bonds.

fig D Straightening your hair changes the arrangement of the hydrogen bonds so the hair curls in a different direction.

Your hair is made of the protein keratin. Some methods of styling the hair actually change the bonds within the protein molecules. Blow drying or straightening your hair breaks the hydrogen bonds and reforms them with the hair curling in a different way temporarily until the hydrogen bonds reform in their original places.

Perming breaks the disulfide bonds between the polypeptide chains and reforms them in a different place. This effect is permanent – your hair will stay styled in that particular way until it is cut.

Protein structure

Proteins can be described by their primary, secondary, tertiary and quaternary structure (see **fig E**).

- The primary structure of a protein is the sequence of amino acids that make up the polypeptide chain held together by peptide bonds.

- The secondary structure of a protein is the arrangement of the polypeptide chain into a regular, repeating structure, held together by hydrogen bonds. One example is the right-handed helix (α-helix), a spiral coil with the peptide bonds forming the backbone and the R groups sticking out in all directions. Another is the β-pleated sheet, in which the polypeptide chain folds into regular pleats held together by hydrogen bonds between the amino and carboxyl ends of different amino acids. Most **fibrous proteins** have this sort of structure. Sometimes there is no regular secondary structure and the polypeptide forms a random coil.

- The tertiary structure is a level of 3D organisation imposed on top of the secondary structure in many proteins. The amino acid chain, including any α-helices and β-pleated sheets, is folded further into complicated shapes. Hydrogen bonds, disulfide bonds and ionic bonds between amino acids hold these 3D shapes in place (see **page 30**). **Globular proteins** are an example of tertiary structures.

- The quaternary structure of a protein is only seen in proteins consisting of several polypeptide chains. The quaternary structure describes the way these separate polypeptide chains fit together in three dimensions. Examples include some very important enzymes and the blood pigment haemoglobin.

The bonds that hold the 3D shapes of proteins together are affected by changes in conditions such as temperature or pH. Even small changes can cause the bonds to break, resulting in the loss of the 3D shape of the protein. We say that the protein is **denatured.** Because the 3D structure of these proteins is important to the way they work, changing conditions inside the body can cause proteins such as enzymes to stop working properly.

Primary structure – the linear sequence of amino acids in a peptide.

Secondary structure – the repeating pattern in the structure of the peptide chains, such as an α-helix or β-pleated sheets.

Tertiary structure – the three-dimensional folding of the secondary structure.

Quaternary structure – the three-dimensional arrangement of more than one tertiary polypeptide.

fig E The 3D structure of proteins.

Fibrous and globular proteins

Fibrous proteins

The complex structures of large protein molecules relate closely to their functions in the body. Fibrous proteins have little or no tertiary structure. They are long, parallel polypeptide chains with occasional cross-linkages that form into fibres. They are insoluble in water and are very tough, which makes them ideally suited to their structural functions within organisms. Fibrous proteins appear in the structure of connective tissue in tendons and the matrix of bones, in the structure of muscles, as the silk of spiders' webs and silkworm cocoons, and as the keratin that makes up hair, nails, horns and feathers.

Collagen is a fibrous protein that gives strength to tendons, ligaments, bones and skin. It is the most common structural protein found in animals – up to 35% of the protein in your body is collagen. Collagen is extremely strong – the fibres have a tensile strength comparable to that of steel. This is due to the unusual structure of the collagen molecule. It is made up of three polypeptide chains, which are each up to 1000 amino acids long. The primary structure of these chains is repeating sequences of glycine with two other amino acids – often proline and hydroxyproline. The three α-chains are arranged in a unique triple helix, held together by a very large number of hydrogen bonds. These collagen molecules, which can be up to several millimetres long, are often found together in fibrils that in turn are held together to form collagen fibres.

Collagen fibres are found combined with the bone tissue, giving it tensile strength rather like the steel rods in reinforced concrete. In the genetic disease osteogenesis imperfecta, the collagen triple helix may not form properly. The bone lacks tensile strength as a result, and it is brittle and breaks very easily.

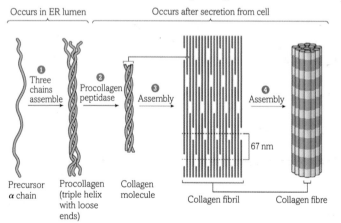

fig F Collagen is a fibrous protein of with an unusual triple helix structure and immense strength

Globular proteins

Globular proteins have complex tertiary and sometimes quaternary structures. They fold into spherical (globular) shapes. The large size of these globular protein molecules affects their behaviour in water.

Because their carboxyl and amino ends give them ionic properties you might expect them to dissolve in water and form a solution. In fact, the molecules are so big that instead they form a colloid. Globular proteins play an important role in holding molecules in position in the cytoplasm. Globular proteins are also important in your immune system – for example, antibodies are globular proteins. Globular proteins form enzymes and some hormones and are involved in maintaining the structure of the cytoplasm (see **Section 1.4.1** for details of proteins as enzymes).

Haemoglobin is one of the best known globular proteins. It is a very large molecule made up of 574 amino acids arranged in four polypeptide chains which are held together by disulfide bonds. Each chain is arranged around an iron-containing haem group. Haemoglobin is a **conjugated protein** as well as a globular protein. It is the iron that enables the haemoglobin to bind and release oxygen molecules, and it is the arrangement of the polypeptide chains that determines how easily the oxygen binds or is released (see **Section 4.3.3**)

Conjugated proteins

Some protein molecules are joined with or conjugated to another molecule called a **prosthetic group**. This structural change usually affects the performance and functions of the molecules. You have already looked at haemoglobin, a large protein with an iron-containing prosthetic group. Chlorophyll, the molecule involved in the capture of light energy in photosynthesis, is another conjugated protein, with a prosthetic group that contains magnesium.

Glycoproteins are proteins with a carbohydrate prosthetic group. The carbohydrate part of the molecule helps them to hold on to a lot of water and also makes it harder for protein-digesting enzymes (**proteases**) to break them down. Lots of lubricants used by the human body – such as mucus and the synovial fluid in the joints – are glycoproteins whose water-holding properties make them slippery and viscous, which reduces friction. This also helps to explain why the mucus produced in the stomach protects the protein walls from digestion.

Lipoproteins are proteins conjugated with lipids and are very important in the transport of cholesterol in the blood. The lipid part of the molecule enables it to combine with the lipid cholesterol. There are two main forms of lipoproteins in your blood – low-density lipoproteins (LDLs) (around 22 nm in diameter) and high-density lipoproteins (HDLs) (around 8–11 nm in diameter). The HDLs contain more protein than LDLs, which is partly why they are denser – proteins are more compact molecules than lipids.

> **Learning tip**
>
> Remember that amino acids are joined together by peptide bonds to form dipeptides and then polypeptides, but the 3D structures of proteins are the result of hydrogen bonds, disulfide bonds and ionic bonds between amino acids within the polypeptide chains.

Did you know?

Testing for protein

To test for the presence of protein, either add 5% (w/v) potassium or sodium hydroxide solution and 1% (w/v) copper sulfate solution, or Biuret reagent which is the two chemicals ready mixed. When the reagent/s are added to a test solution, a purple colour indicates the presence of protein.

Questions

1 Explain how the order of amino acids in a protein affects the structure of the whole protein.

2 Hydrogen bonds are weaker than disulfide bonds and ionic salt bridges, but they play a much bigger role in maintaining protein structure. Why is this?

3 The body uses many resources to maintain a relatively constant internal environment. With reference to proteins, explain why constant internal conditions are so important.

Key definitions

Amino acids are the building blocks of proteins, consisting of an amino group ($-NH_2$) and a carboxyl group ($-COOH$) attached to a carbon atom and an R group that varies between amino acids.

A **peptide bond** is the bond formed by condensation reactions between amino acids.

A **dipeptide** is two amino acids joined by a peptide bond.

A **polypeptide** is a long chain of amino acids joined by peptide bonds.

Fibrous proteins are proteins that have long, parallel polypeptide chains with occasional cross-linkages that form into fibres, but with little tertiary structure.

A **disulfide bond** is a strong covalent bond formed as a result of an oxidation reaction between sulfur groups in cysteine or methionine molecules, which are close together in the structure of a polypeptide.

Globular proteins are large proteins with complex tertiary and sometimes quaternary structures, folded into spherical (globular) shapes.

Haemoglobin is a large conjugated protein involved in transporting oxygen in the blood, and gives the erythrocytes their red colour.

Collagen is a strong fibrous protein with a triple helix structure.

Denaturation is the loss of the 3D shape of a protein, e.g. as a result of changes in temperature or pH.

A **prosthetic group** is the molecule that is incorporated in a conjugated protein.

A **glycoprotein** is a protein with a carbohydrate prosthetic group.

A **protease** is a protein-digesting enzyme.

A **lipoprotein** is a protein with a lipid prosthetic group.

TREHALOSE – A SUGAR FOR DRY EYES?

Biological molecules have an amazing number of different roles in living organisms, including some you would not expect. In this activity you will discover how current research, which shows that disaccharide trehalose can protect proteins from damage in stressful conditions, is being used to make dry eyes more comfortable – and possibly protect the brain from the damage that can result from ageing.

TREHALOSE: AN INTRIGUING DISACCHARIDE WITH POTENTIAL FOR MEDICAL APPLICATION IN OPHTHALMOLOGY

Jacques Luyckx and Christophe Baudouin

Abstract

Trehalose is a naturally occurring disaccharide comprised of two molecules of glucose. The sugar is widespread in many species of plants and animals, where its function appears to be to protect cells against desiccation, but it is not found in mammals. Trehalose has the ability to protect cellular membranes and labile proteins against damage and denaturation as a result of desiccation and oxidative stress. Trehalose appears to be the most effective sugar for protection against desiccation. Although the exact mechanism by which trehalose protects labile macromolecules and lipid membranes is unknown, credible hypotheses do exist. As well as being used in large quantities in the food industry, trehalose is used in the biopharmaceutical preservation of labile protein drugs and in the cryopreservation of human cells. Trehalose is under investigation for a number of medical applications, including the treatment of Huntington's chorea and Alzheimer's disease. Recent studies have shown that trehalose can also prevent damage to mammalian eyes caused by desiccation and oxidative insult. These unique properties of trehalose have thus prompted its investigation as a component in treatment for dry eye syndrome. This interesting and unique disaccharide appears to have properties which may be exploited in ophthalmology and other disease states.

Trehalose, a naturally occurring alpha-linked disaccharide formed of two molecules of glucose (**fig A**) … is synthesized by many living organisms, including insects, plants, fungi, and micro-organisms as a response to prolonged periods of desiccation. This very useful property, known as anhydrobiosis, confers on an organism the ability to survive almost complete dehydration for prolonged periods and subsequently reanimate.

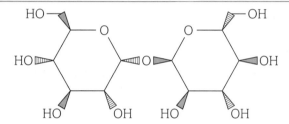

fig A Structure of trehalose. Registry number: 99-20-7; Molar mass: 342.296 g/mol (anhydrous); 378.33 g/mol (dihydrate); molecular structure: α-D-glucopyranosyl α-D-glucopyranoside (α,α-trehalose).

References

1   Iturriaga G, Suárez R, Nova-Franco B. Trehalose metabolism: From osmoprotection to signaling. *Int J Mol Sci*. 2009; 10:3793–3810. […]

8   Elbein AD, Pan YT, Pastuszak I, Carroll D. New insights on trehalose: A multifunctional molecule. *Glycobiology*. 2003; 13:17R–27R.

11 Jain NK, Roy I. Effect of trehalose on protein structure. *Protein Sci*. 2009; 18:24–36. […]

20 Matsuo T. Trehalose protects corneal epithelial cells from death by drying. *Br J Ophthalmol*. 2001; 85:610–612. […]

30 Matsuo T, Tsuchida Y, Morimoto N. Trehalose eye drops in the treatment of dry eye syndrome. *Ophthalmology*. 2002; 109:2024–2029.

31 Matsuo T. Trehalose versus hyaluronan or cellulose in eyedrops for the treatment of dry eye. *Jpn J Ophthalmol*. 2004; 48:321–327.

Where else will I encounter these themes?

1.1   1.2 YOU ARE HERE   1.3   1.4   2.1   2.2   2.3   2.

Let us start by considering the nature of the writing in this article:

1. This extract comes from a paper published in *Clinical Ophthalmology*, an online journal. Think about the type of writing being used and the audience it is intended for as you try and answer the following questions:

   a. What aspects of this writing tell you it is from a scientific paper rather than a general interest article in a magazine?

   b. Choose two words used in the article that you are not familiar with. Find out what they mean and suggest why they have been used by the author of the article.

   c. How do you think these ideas about trehalose and the way it may be used to help human health might be presented in a newspaper or on the BBC website? Have a go at writing an article for a public interest website yourself.

   d. If trehalose can really help protect people's sight and prevent brain diseases such as Huntington's and Alzheimer's this would make a big difference to people's lives. Notice how cautious the author is. Why are scientific papers so measured in the way they report things?

Think about the level of scientific detail that is suitable for your expected audience. How will you ensure your article is eye-catching and interesting?

Now you are going to think about the science in the article. You will be surprised how much you know already, but if you choose to do so, you can return to these questions later in your course.

2. What do you know about the chemical nature of trehalose from the article?

3. Desiccation (drying out) is a major problem for living organisms. Suggest reasons why drying out is so hard to survive.

4. Scientists think that trehalose protects both lipid membranes and certain proteins from damage, both from drying out and oxidation. Explain why it is so important biologically to protect cell membranes and protein structures.

Think about the chemistry of biological molecules you have learned already and use it to help you understand how trehalose works.

Activity

fig B *Selaginella lepidophylla* is a resurrection plant – it can withstand almost complete dehydration and recover within about 24 hours, thanks to high levels of trehalose in the plant cells.

You can refer to the full version of this paper, to the references listed at the end, to online encyclopedias, to other scientific papers and to books. In each case, judge the reliability of your source before you use it.

Which aspect of trehalose would you like to know more about? The way it prevents desiccation in many groups of organisms? The way it can protect human eyes from damage? The evidence that it could help reduce brain diseases in people?

Choose the area that interests you most and use as many resources as you can to produce a 3-minute presentation about that aspect of trehalose biology. Find interesting images and list all the references to help your colleagues decide if they can rely on the information you present.

● From the following journal article:
  'Trehalose: an intriguing disaccharide with potential for medical application in ophthalmology.' *Clinical Ophthalmology* (Auckland, NZ) 5 (2011): 577.

Exam-style questions

1 Which statement best describes the structure or role of these biological molecules?

(a) Dissacharides can be split by

   A hydrolysis of glycosidic bonds

   B condensation of glycosidic bonds

   C hydrolysis of ester bonds

   D condensation of ester bonds [1]

(b) Amylose is an example of a

   A monosaccharide

   B disaccharide

   C polysaccharide

   D trisaccharide [1]

(c) The role of starch is to

   A be a source of energy to plants

   B store energy in all living organisms

   C store energy in plants

   D store energy in animals [1]

(d) Proteins are polymers of amino acids joined by peptide bonds formed between the

   A R groups

   B R group and the amino group

   C R group and the carboxyl group

   D carboxyl group and the amino group [1]

(e) The three-dimensional structure of a protein is held together by

   A peptide, hydrogen and ionic bonds

   B hydrogen, ester and ionic bonds

   C disulfide bridges and ester bonds

   D disulfide bridges, hydrogen and ionic bonds [1]

(f) DNA consists of mononucleotides joined togther by bonds between

   A two pentose sugars

   B one ribose sugar and one phosphate group

   C one deoxyribose sugar and one phosphate group

   D two phosphate groups [1]

(g) Water is described as a polar molecule because it has a

   A positively charged hydrogen end and a negatively charged oxygen end

   B positively charged hydrogen end and a positively charged oxygen end

   C negatively charged hydrogen end and a negatively charged oxygen end

   D negatively charged hydrogen end and a positively charged oxygen end [1]

[Total: 7]

2 Fill in this table to show the components and bonding within each carbohydrate.

	Lactose	Maltose	Amylose
Component monosaccharides			
Bonds between monosaccharides			

[3]

[Total: 3]

3 A disaccharide can be hydrolysed to its two monosaccharides. Explain the term **hydrolysis**. [2]

[Total: 2]

4 Read through the following account on lipids, then write on the dotted lines the most appropriate words to complete the account.

Lipids are insoluble in water because they are

....

A triglyceride is one type of lipid. A triglyceride consists of

one .... molecule with three

.... molecules joined to it by

.... bonds. Triglycerides have

important roles in living organisms, including waterproofing

and .... [5]

[Total: 5]

5 (a) Draw a diagram to show the structure of a phospholipid. Use the symbols shown for each component in your diagram.

Glycerol: [rectangle] Phosphate group: [circle]

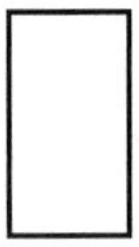

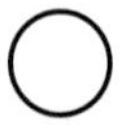

Fatty acid: [bar] Ester bond: —— [3]

(b) The presence of a phosphate makes part of the molecule hydrophilic. Explain what is meant by the term **hydrophilic**. [1]

(c) Describe the role of phospholipids in the cell surface (plasma) membrane. [2]

[Total: 6]

6 (a) Draw a triglyceride. You may use any component more than once.

[rectangle] glycerol   〰〰〰 fatty acid   —— ester bond

(b) There are four statements about triglycerides given below. If the statement is correct, put a tick (✓) in the box to the right of that statement. If the statement is incorrect, put a cross (✗) in the box to the right of the statement.

Statement	Tick (✓) or cross (✗)
Triglycerides are building blocks of polysaccharides	
Triglycerides can contain a small amount of nitrogen	
Triglycerides can be modified into phospholipids	
Triglycerides release water during hydrolysis	

[4]

(c) Fatty acids can be either saturated or unsaturated. Explain what is meant by the term **saturated** fatty acid. [1]

[Total: 5]

7 Describe the structure of an amino acid. [2]

[Total: 2]

8 (a) Insulin and collagen are both proteins that have a primary structure made up of amino acids joined together by peptide bonds.
   (i) Explain what is meant by the term **primary structure** of a protein. [1]
   (ii) Name the type of reaction that occurs when a peptide bond is broken causing a dipeptide to split into two amino acids. [1]

(b) Insulin and collagen both contain the amino acids glycine and serine. The diagram below shows a dipeptide formed from these two amino acids. Complete the diagram to show the structure of serine when the peptide bond breaks.

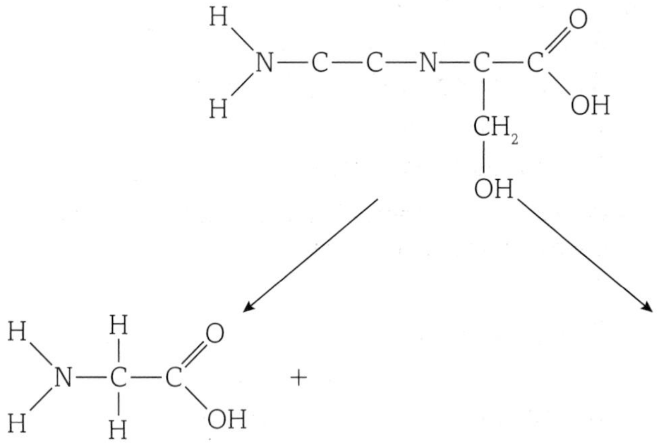

Glycine   +   Serine

[1]

[Total: 3]

TOPIC 1
Biological molecules

1.3 ▷ Biological molecules 2

Introduction

In an air-conditioned room near Cambridge, ranks of machines at the Wellcome Trust Sanger Institute sequence the genetic material of thousands of anonymous people, of disease-causing bacteria and of cancers that mutate and grow in spite of chemotherapy. Without noise, without drama, the secrets of life itself are revealed in bars of light as the ever-developing technology identifies the sequences of bases that make up the DNA code. The first complete sequence of the human genome took years of work by scientists in many countries. Now it takes days to complete a human genome and less than 24 hours to sequence the genetic material of a bacterium. The expertise of scientists is still needed to interpret and use the information, which is being produced 24 hours a day. The potential benefits from this rapidly developing area of science, where biology, medicine and computers come together, are almost limitless.

In this chapter you will be studying the nucleotides and some of the molecules in which they play a key role, including DNA.

Adenosine triphosphate (ATP) is the molecule that acts as the universal energy supply in cells of every type. You will look at the structure of the molecule and how this is related to its role in cells. You will be referring to ATP in almost every aspect of your biology studies.

Nucleic acids or polynucleotides are the information molecules of the cell. You will discover the structure of deoxyribonucleic acid (DNA) and crack the code by which it carries the information needed to build an entire new organism.

You will learn about the different types of ribonucleic acid (RNA) and how they work together with the DNA to translate the genetic code into the phenotype of the cell or organism through protein synthesis. You also will build up a model of mutation and see how changes in the genetic code itself can result in changes, which may benefit or damage an organism.

All the maths you need

- Recognise and make use of appropriate units in calculations (*e.g. nanometres*)
- Use ratios (*e.g. representing the relationships between atoms in an ion or molecule*)

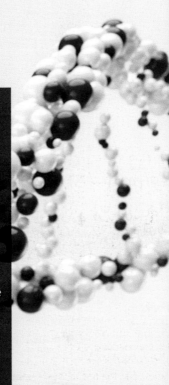

What have I studied before?

- DNA is a polymer made up of two strands that form a double helix
- DNA is the genetic material of the cell
- The genome is the entire genetic material of the cell
- Our increasing understanding of the human genome is potentially very important in medicine
- The impact of genome sequencing on how we classify organisms
- Mutations cause variants, most of which have no effect on the phenotype, some of which influence the phenotype and some of which determine the phenotype of the organism. These may be positive or negative in their effect
- Cellular respiration is an exothermic reaction that produces ATP

What will I study later?

- The importance of the enzymes formed during protein synthesis
- The site of the genetic material in the nucleus of the cell
- What happens to the genetic material during mitotic and meiotic cell division
- What happens to the DNA in chromosome mutations and how they can affect the phenotype of the resulting individual
- The impact of DNA sequencing on classification (A level)
- The importance of mutation and genetic variation in evolution by natural selection (A level)
- Biodiversity within the gene pool of a population (A level)
- The role of ATP in cellular respiration during glycolysis and the Krebs cycle (A level)
- The synthesis of ATP by chemiosmosis (A level)
- The role of ATP in the light-dependent and the light-independent stages of photosynthesis (A level)
- Gene technology (A level)

What will I study in this chapter?

- The structure of nucleotides
- The structure of ATP related to its function as the universal energy supplier to cells
- The structure of the DNA molecule including the double helix
- The story of how the double helix structure of DNA was discovered
- How the DNA code works and the experimental evidence used to prove it
- Protein synthesis and the roles of the different types of RNA in the process
- Different types of gene mutations and the effect they have on the amino acid sequences of the proteins that are formed

By the end of this section, you should be able to...

● outline the structure of nucleotides, both purines and pyrimidines

● relate the structure and properties of ATP to its function in the cell

Nucleotides

Nucleotides are key molecules in biology. They provide the energy currency of cells in the form of **adenosine triphosphate**, usually referred to as **ATP**. They also provide the building blocks for the mechanism of inheritance in the form of DNA – deoxyribonucleic acid – and RNA – ribonucleic acid.

Each nucleotide has three parts – a 5-carbon pentose sugar, a nitrogen-containing base and a phosphate group. The pentose sugar in RNA is ribose, and in DNA is deoxyribose. Deoxyribose, as its name suggests, contains one fewer oxygen atom than ribose (see **fig A**).

The most common types of nucleotides have either a **purine base**, which has two nitrogen-containing rings, or a **pyrimidine base**, which has only one. Both purines and pyrimidines are weak bases. The most common purines are **adenine** (A) and **guanine** (G) and the most common pyrimidines are **cytosine** (C), **thymine** (T) and **uracil** (U).

A phosphate group ($-PO_4^{3-}$) is the third component of a nucleotide. Inorganic phosphate ions are present in the cytoplasm of every cell (see **Section 1.1.1** for more about inorganic ions). It is as a result of this phosphate group that the nucleotides are acidic molecules and carry a negative charge.

The sugar, the base and the phosphate group are joined together by condensation reactions, with the elimination of two water molecules, to form a nucleotide (see **fig A**).

ATP

Cells are chemical factories, with many reactions continually taking place within the cytoplasm and organelles (see **Sections 2.1.3** and **2.1.5** on cell structures and **Book 2 Chapter 5.1** on cellular respiration). In these reactions, chemical bonds are constantly being broken and energy is always needed to break the bonds. Each cell needs a constantly available and immediately accessible supply of energy to support a multitude of different reactions.

One molecule seems to be the universal energy supplier in cells – adenosine triphosphate (ATP). ATP is found in all living organisms in exactly the same form. Anything which interferes with the production or breakdown of ATP is fatal to the cell and ultimately destroys the whole multicellular organism.

ATP is a nucleotide with three phosphate groups attached (see **fig B**). It is the potential energy in the phosphate bonds, that is made available to cells for use in breaking bonds in chemical reactions.

Fig B shows the structure of ATP. When energy is needed in the cell, the third phosphate bond in the molecule is broken in a hydrolysis reaction. This is catalysed by the enzyme **ATPase**. The products of the reaction are **adenosine diphosphate (ADP)**, another nucleotide, and a free inorganic phosphate group (P_i). One phosphate bond is broken as the ATP is split – this uses energy. Two further bonds are made to produce the ADP and the stable phosphate group and this releases the energy that is needed to drive other reactions. About 34 kJ of energy are released per mole of ATP hydrolysed. Some of this energy is always lost to the system as heat, but the rest is used for any energy-requiring biological activity in the cell such as building up new molecules, active transport (**see Section 4.1.4**), nerve impulses or muscle contraction.

fig A The structure of a nucleotide. The properties of the nucleotide molecule are crucial to the roles of ATP, DNA and RNA.

(a)

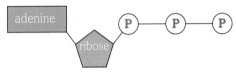

adenine

OH OH

ribose   3 × phosphates

(b)

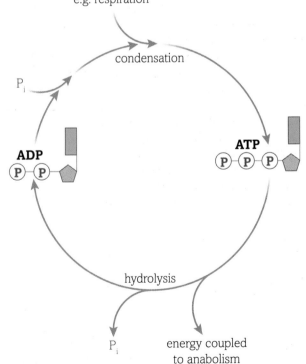

fig B The structure of ATP.

The breakdown of ATP into ADP is a reversible reaction. ATP can be synthesised from ADP and a phosphate group in a reaction that requires an input of energy (30.5 kJ per mole of ATP produced). ATPase catalyses this reaction. The direction of the reaction depends on conditions in the cell. The energy needed to drive the synthesis of ATP usually comes from breakdown reactions or from **reduction/oxidation (redox) reactions**. As a result, the ATP molecule provides an immediate supply of energy, ready for use when needed.

coupled to catabolism, e.g. respiration

condensation

P_i

ADP

ATP

hydrolysis

P_i energy coupled to anabolism

fig C The energy released in catabolic reactions drives the production of ATP. When needed, this energy can then be used to drive synthesis reactions in the cell.

Did you know?

Cyanide and ATP

Cyanide is a well-known poison that smells of bitter almonds. It is fatal because the poison blocks part of the process of cellular respiration producing ATP. Without ATP the cells of the body stop working. The muscles go into spasm and the victim cannot breathe, which rapidly results in death.

Questions

1 Describe the structure of a nucleotide.

2 ATP is regarded as the universal energy supply molecule. Why is this and how does its structure relate to its role in cells?

Key definitions

Nucleotides are molecules with three parts – a 5-carbon pentose sugar, a nitrogen-containing base and a phosphate group – joined by condensation reactions.

Adenosine triphosphate (ATP) is a nucleotide that acts as the universal energy supply molecule in cells. It is made up of the base adenine, the pentose sugar ribose and three phosphate groups.

A **purine base** is a base found in nucleotides that has two nitrogen-containing rings.

A **pyrimidine base** is a base found in nucleotides that has one nitrogen-containing ring.

Adenine is a purine base found in DNA and RNA.

Guanine is a purine base found in DNA and RNA.

Cytosine is a pyrimidine base found in DNA and RNA.

Thymine is a pyrimidine base found in DNA.

Uracil is a pyrimidine base found in RNA.

ATPase is an enzyme that catalyses the formation and the breakdown of ATP, depending on conditions.

Adenosine diphosphate (ADP) is a nucleotide formed when ATP loses a phosphate group and provides energy to drive reactions in the cell.

Reduction/oxidation (redox) reactions are reactions in which one reactant loses electrons (is oxidised) and another gains electrons (is reduced).

Reproduction is one of seven key processes in living organisms. If the individuals in a species do not reproduce, then that species will die out. Multicellular organisms also need to grow, and to replace worn-out cells. Within every cell is a set of instructions for the assembling of new cells. These can be used both to form offspring and to produce identical cells for growth. Over the last 75 years or so scientists have made enormous strides towards understanding the form of these instructions – the genetic code. In unravelling the secrets of the genetic code, people have come closer than ever before to understanding the mystery of life itself.

Nucleic acids, also known as polynucleotides, are the information molecules of the cell. They are polymers, made of many nucleotide monomer units. They carry all the information needed to form new cells. The information takes the form of a code in the molecules of DNA – deoxyribonucleic acid (see **fig C**). Parts of the code are copied into messenger RNA (mRNA) and used to direct the production of the proteins that build the cell and control its actions. In eukaryotic cells, the genetic information is stored in chromosomes in the nucleus (see **Section 2.1.3**), but in prokaryotes a single length of DNA is found floating freely in the cytoplasm (see **Section 2.2.1**).

Building the polynucleotides

Nucleic acids are chains of nucleotides linked together by condensation reactions that produce **phosphodiester bonds** between the sugar on one nucleotide and the phosphate group of the next nucleotide. These nucleic acids can be millions of nucleotide units long. Both DNA and RNA have this sugar–phosphate backbone. Because the sugar of one nucleotide bonds to the phosphate group of the next nucleotide, polynucleotides always have a hydroxyl group at one end and a phosphate group at the other. This structural feature is important in the role of the nucleic acids in the cell. Long chains of nucleotides containing the bases C, G, A and T join together to form DNA. Chains of nucleotides containing C, G, A and U make RNA. Knowledge of how these units join together, and the three-dimensional (3D) structures in DNA in particular, is the basis of our understanding of molecular genetics.

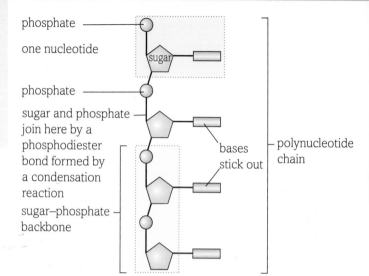

fig A A polynucleotide strand like this makes up the basic structure of both DNA and RNA.

RNA molecules form single polynucleotide strands that can fold into complex shapes, held in place by hydrogen bonds, or remain as long thread-like molecules. DNA molecules consist of two polynucleotide strands twisted around each other. The sugars and phosphates form the backbone of the molecule and, pointing inwards from the two sugar–phosphate backbones, are the bases, which pair up in specific ways. A purine base always pairs with a pyrimidine base – in DNA, adenine pairs with thymine and cytosine with guanine. This results in the famous DNA double helix, a massive molecule that resembles a spiral staircase.

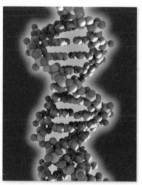

fig B The double helix structure of a DNA molecule is not just an iconic image of science – it is vital to the role of DNA in cells.

The two strands of the DNA double helix are held together by hydrogen bonds between the complementary base pairs (see **fig C**). These hydrogen bonds form between the amino and the carbonyl groups of the purine and pyrimidine bases on the opposite strands. There are three hydrogen bonds between C and G but only two between A and T. There are 10 base pairs for each complete twist of the helix. The two strands are known as the 5' (5 prime) and 3' (3 prime) strand, named according to the number of the carbon atoms in the pentose sugar to which the phosphate group is attached in the first nucleotide of the chain. It is the phosphate that is free at the 5' end of the 5' carbon, and it is the free –OH group that is attached to the 3' carbon on the 3' end. As you will see, these features of the structure of DNA and RNA are crucial to the way the molecules function within cells.

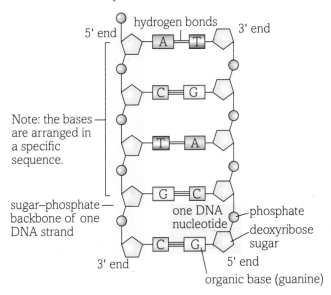

Note: the bases are arranged in a specific sequence.

sugar–phosphate backbone of one DNA strand

one DNA nucleotide

phosphate

deoxyribose sugar

organic base (guanine)

Purines   **Pyrimidines**

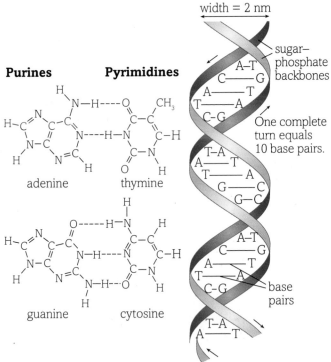

adenine   thymine

guanine   cytosine

width = 2 nm

sugar–phosphate backbones

One complete turn equals 10 base pairs.

base pairs

The two strands are antiparallel – one runs in one direction and the other in the opposite direction.

fig C The double helix structure of DNA depends on the hydrogen bonds that form between the base pairs.

Learning tip

Make sure you use the terms nucleotide and nucleic acid correctly – do not muddle them up.

Did you know?

Sequencing the genome

From the late twentieth century onwards scientists from around the world collaborated in the Human Genome Project. This was an ambitious project that set out to identify all of the genes in the human chromosomes and to sequence the 3 billion base pairs which make up the human DNA. The scientists worked on DNA from anonymous donors and showed that every individual has at least 99.9% of their DNA in common. Leaps in technology meant the project finished ahead of the expected date, but it still took almost 13 years.

In 2008 a new project began – the 1000 Genomes Project. This time, scientists analysed the DNA of 1092 people from all around the world, to gain information about differences in our DNA that can, amongst many things, have an impact on the diseases that may affect us. The 1000 Genomes Project took 6 years.

The 10K (ten thousand) Genomes Project got under way in 2013. This project is sequencing the genomes of 10 000 people from around the world with rare genetic diseases and cancers. The whole 10K Genomes Project is expected to take only 3 years, thanks to the immense improvements in sequencing technology, which mean that processes that once took weeks and months now take hours and days. It should greatly increase our understanding, diagnosis and even treatment of rare genetic conditions.

fig D DNA sequencing in progress at the Wellcome Trust Sanger Institute.

Questions

1 What is a nucleotide monomer unit and which constituent parts are found in both DNA and RNA?

2 (a) Explain how complementary base pairing and hydrogen bonding are responsible for the structure of DNA.

   (b) Look carefully at the structural formulae of the purine bases and the pyrimidine bases of the DNA molecule. Suggest reasons why the pairs of bases always involve one purine and one pyrimidine base, never two purines or two pyrimidines.

Key definitions

Nucleic acids are polymers made up of many nucleotide monomer units that carry all the information needed to form new cells.

A **phosphodiester bond** is the bond formed between the phosphate group of one nucleotide and the sugar of the next nucleotide in a condensation reaction.

A **genome** is the entire genetic material of an organism.

One of the most important features of the DNA molecule is that it can **replicate**, or copy itself, exactly. This is the characteristic above all others, that means it can pass on genetic information from one cell or generation to another.

Uncovering the mechanism of replication

After Watson and Crick had produced their double helix model for the structure of DNA, it took scientists some years to work out exactly how the molecule replicates itself.

There were two main ideas about how replication happens: conservative and semiconservative replication. In the conservative replication model, the original double helix remained intact and in some way instructed the formation of a new, identical double helix, made up entirely of new material. ^{14}N

The semiconservative replication model assumed that the DNA 'unzipped' and new nucleotides aligned along each strand. Each new double helix contained one strand of the original DNA and one strand made up of new material. This was the Watson and Crick hypothesis – the double helix would unzip along the hydrogen bonds in their structural model, allowing semiconservative replication to take place. It took a classic piece of practical investigation to settle the argument.

Experimental evidence

As the result of a very elegant set of experiments carried out by Matthew Meselson (1930–) and Franklin Stahl (1929–) at the California Institute of Technology in the late 1950s, semiconservative replication became the accepted model of DNA replication.

- They grew several generations of the gut bacteria *Escherichia coli* (*E. coli*) in a medium where their only source of nitrogen was the isotope ^{15}N from ^{15}NH$_4$Cl. Atoms of ^{15}N are denser than those of the isotope usually found, ^{14}N. The bacteria grown on this medium took up the isotope to make the cell chemicals, including proteins and DNA. After several generations, the entire bacterial DNA was labelled with ^{15}N ('heavy' nitrogen).

- They moved the bacteria to a medium containing normal ^{14}NH$_4$Cl as their only nitrogen source, and measured the density of their DNA as they reproduced.

- Meselson and Stahl predicted that if DNA reproduced by conservative replication, some of the DNA would have the density expected if it contained nothing but ^{15}N (the original strands), and some of it would have the density expected if it contained nothing but ^{14}N (the new strands). However, if DNA reproduced by semiconservative replication, then all of the DNA would have the same density, half-way between that of ^{15}N- and ^{14}N-containing DNA.

- They found that all DNA had the same density, half-way between that of ^{15}N- and ^{14}N-containing DNA – and so DNA must replicate semiconservatively.

Conservative replication, where the double helix remains intact and new strands form on the outside, would give:

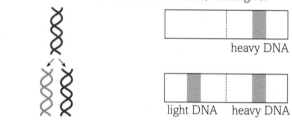

Replicates in medium containing only light nitrogen.

Half of the DNA molecules have 2 light strands and half have 2 heavy strands.

Semiconservative replication, where the double helix unzips and each strand replicates to produce a second, new strand, would give:

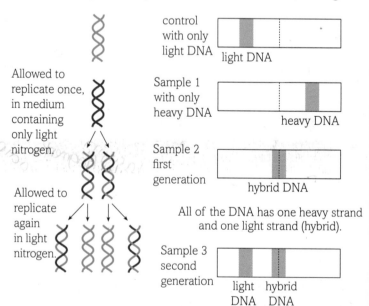

Half of the DNA molecules have light DNA and half are hybrid with one light and one heavy strand.

fig A The results of these experiments by Meselson and Stahl put an end to the theory of conservative replication of DNA.

How DNA makes copies of itself

A careful look at the process of the semiconservative replication of DNA shows clearly the importance of the structure and properties of the DNA molecule to its role as the genetic material of the cell.

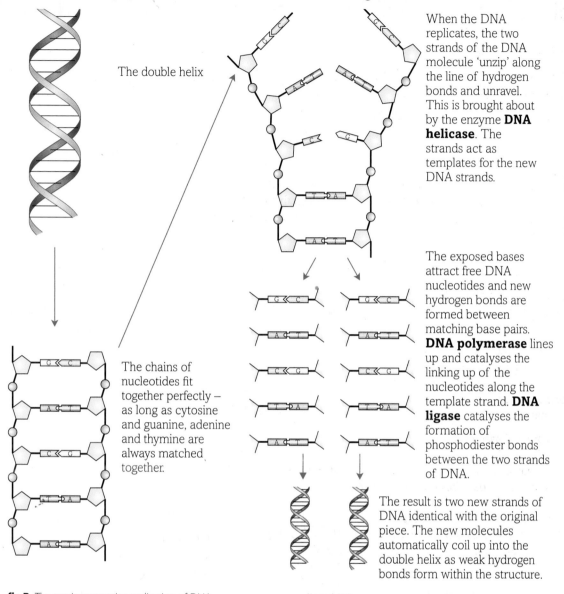

The double helix

The chains of nucleotides fit together perfectly – as long as cytosine and guanine, adenine and thymine are always matched together.

When the DNA replicates, the two strands of the DNA molecule 'unzip' along the line of hydrogen bonds and unravel. This is brought about by the enzyme **DNA helicase**. The strands act as templates for the new DNA strands.

The exposed bases attract free DNA nucleotides and new hydrogen bonds are formed between matching base pairs. **DNA polymerase** lines up and catalyses the linking up of the nucleotides along the template strand. **DNA ligase** catalyses the formation of phosphodiester bonds between the two strands of DNA.

The result is two new strands of DNA identical with the original piece. The new molecules automatically coil up into the double helix as weak hydrogen bonds form within the structure.

fig B The semiconservative replication of DNA.

Learning tip

Make sure you are clear about the difference between conservative and semiconservative models of DNA replication and can explain how the evidence supports the second model.

Questions

1 Make a flow diagram to describe the replication of DNA.

2 How did the work of Meselson and Stahl destroy support for the model of the conservative replication of DNA?

Key definitions

When a DNA molecule **replicates**, it copies itself exactly.

DNA helicase is an enzyme involved in DNA replication that unzips the two strands of the DNA molecule.

DNA polymerase is an enzyme involved in DNA replication that lines up and catalyses the linking up of the nucleotides along the template strand.

DNA ligase is an enzyme involved in DNA replication that catalyses the formation of phosphodiester bonds between the two strands of DNA.

1.3 ▶ 4 ▶ The genetic code

By the end of this section, you should be able to...

● define a gene as a sequence of bases on a DNA molecule coding for a sequence of amino acids in a polypeptide chain

● explain the nature of the genetic code including triplets, codons, the degenerate and non-overlapping nature of the code and that not all of the genome codes for proteins

We know that DNA has a double helix structure and can replicate itself exactly. But how does DNA act as the genetic material and carry the information needed to make new cells and whole new organisms? The key is the link between DNA and proteins. DNA controls protein synthesis and so the DNA instructions control not only how the cell is built, but also how it works.

Proteins are made up of amino acids. Using the DNA code, the 20 naturally occurring amino acids are joined together in countless combinations to make an almost infinite variety of proteins. This process of **translation** happens on the surface of the **ribosomes** (see **Section 1.3.5** on protein synthesis).

What is the genetic code?

In the DNA double helix, the components that vary are the bases. So scientists guessed that it was the arrangement of the bases that carries the genetic code – but how? There are only four bases, so if one base coded for one amino acid there could be only four amino acids. Even two bases do not give enough amino acids – the possible arrangements of four bases into groups of two is $4 \times 4 = 16$. However, a **triplet code** of three bases gives $4 \times 4 \times 4 = 64$ possible combinations – more than enough for the 20 amino acids that are coded for.

Cracking the code

The genetic code is based on **genes**. We can define a gene as a sequence of bases on a DNA molecule coding for a sequence of amino acids in a polypeptide chain, that affect a characteristic in the phenotype of the organism. By the early 1960s it had been proved that a triplet code of bases was the cornerstone of the genetic code. Each sequence of three bases along a strand of DNA codes for something very specific. Most code for a particular amino acid, but some triplets signal the beginning or the end of one particular amino acid sequence.

A sequence of three bases on the DNA or RNA is known as a **codon**. The codons of the DNA are difficult to work out because the molecule is so large, so most of the work was done on the codons of the smaller molecule mRNA. This mRNA is formed as a **complementary strand** to the DNA, so it is like a reverse image of the original base sequence. Once we know the RNA sequence, we can work out the DNA sequence because of the way bases always pair: T/U with A, and G with C. Sequencing tasks like this have become much easier in the twenty-first century as technology has advanced.

The result of all this work is a sort of dictionary of the genetic code (see **tables A and B**). Much of the original work, done in the 1960s, used the gut bacteria *E. coli*, but all the studies done since suggest that the genetic code is identical throughout the living world.

Large parts of the DNA do not code for proteins. Scientists think the non-coding DNA sequences are very important – 98% of the human DNA is non-coding. They know they are involved in regulating the protein-coding sequences – effectively turning genes on or off. Many organisms have similar non-coding sequences, which suggests they are useful, but in many cases we still do not know exactly what they do.

In the 2% of the human DNA that codes for proteins, some codons code for a particular amino acid, while others code for the beginning or the ending of a particular amino acid sequence. We now know that the genetic code is not only a triplet code, it is non-overlapping and degenerate as well.

Did you know?

DNA code-breakers

The first breakthrough in decoding the genetic code came in 1961 when M.W. Nirenberg (1927–2010) and J.H. Matthaei (1929–) in the United States prepared artificial mRNA where all the bases were uracil. They added their polyU – chains reading UUUUUUUUUUUU ... – to all the other ingredients needed for protein synthesis (ribosomes, tRNAs, amino acids, etc). When they analysed the polypeptides made, they found chains of a single type of amino acid, phenylalanine. UUU appeared to be the mRNA codon for phenylalanine. So the DNA codon would be AAA. The scientists soon showed that CCC codes for proline and AAA for lysine. Evidence for the triplet code – three non-overlapping bases with some degeneracy – built up swiftly from this early work. It was also shown that the minimum length of artificial mRNA that would bind to a ribosome was three bases long – a single codon. This would then bind with the corresponding tRNA. From this point on it was a case of careful and precise work to identify all of the codons and their corresponding amino acids.

second letter of the codon

		A		G		T		C		
A	AAA AAG	phenylalanine	AGA AGG	serine	ATA ATG	tyrosine	ACA ACG	cysteine		**A** **G**
	AAT AAC	leucine	AGT AGC		**ATT** **ATC**	stop codon stop codon	**ACT** ACC	stop codon tryptophan		**T** **C**
G	GAA GAG GAT GAC	leucine	GGA GGG GGT GGC	proline	GTA GTG	histidine	GCA GCG GCT GCC	arginine		**A** **G**
					GTT GTC	glutamine				**T** **C**
T	TAA TAG TAT	isoleucine	TGA TGG TGT TGC	threonine	TTA TTG	asparagine	TCA TCG	serine		**A** **G**
	TAC	methionine; start codon			TTT TTC	lysine	TCT TCC	arginine		**T** **C**
C	CAA CAG CAT CAC	valine	CGA CGG CGT CGC	alanine	CTA CTG	aspartic acid	CCA CCG CCT CCC	glycine		**A** **G**
					CTT CTC	glutamic acid				**T** **C**

(right axis: third letter of the codon; left axis: first letter of the codon)

table A The triplet code that underpins all work on genetics: the DNA code.

second letter of the codon

		U		C		A		G		
U	UUU UUC	phenylalanine	UCU UCC	serine	UAU UAC	tyrosine	UGU UGC	cysteine		**U** **C**
	UUA UUG	leucine	UCA UCG		**UAA** **UAG**	stop codon stop codon	**UGA** UGG	stop codon tryptophan		**A** **G**
C	CUU CUC CUA CUG	leucine	CCU CCC CCA CCG	proline	CAU CAC	histidine	CGU CGC CGA CGG	arginine		**U** **C** **A** **G**
					CAA CAG	glutamine				
A	AUU AUC	isoleucine	ACU ACC	threonine	AAU AAC	asparagine	AGU AGC	serine		**U** **C**
	AUA **AUG**	methionine; start codon	ACA ACG		AAA AAG	lysine	AGA AGG	arginine		**A** **G**
G	GUU GUC	valine	GCU GCC	alanine	GAU GAC	aspartic acid	GGU GGC	glycine		**U** **C**
	GUA GUG		GCA GCG		GAA GAG	glutamic acid	GGA GGG			**A** **G**

(right axis: third letter of the codon; left axis: first letter of the codon)

table B The triplet code that underpins all work on genetics: the RNA code for the same amino acids.

A non-overlapping code...

Once scientists had worked out that the genetic code was based on triplets of DNA bases, they wanted to find out how the code was read. Do the triplets of bases follow each other along the DNA strand like beads on a necklace, or do they overlap? For example, the mRNA sequence UUUAGC could code for two amino acids, phenylalanine (UUU) and serine (AGC). On the other hand, if the code overlaps, it could code for four: phenylalanine (UUU), leucine (UUA), a nonsense or stop codon (UAG) and serine (AGC).

An overlapping code would be very economical – relatively short lengths of DNA could carry the instructions for many different proteins. However, it would also be very limiting, because the amino acids that could be coded for side by side would be limited. In the example given, only leucine out of the 20 available amino acids could ever follow phenylalanine, because only leucine has an mRNA codon starting with UU–.

Scientists rely on experimental observations to help decide whether the genetic code is overlapping or not. If a codon consists of three nucleotides and is completely overlapping, and a single nucleotide is altered by a point mutation, then three amino acids will be affected by that single change. If the code is only partly overlapping, then a single point mutation would result in two affected amino acids. But if the codons do not overlap at all, then a change in a single nucleotide mutation would affect only one amino acid, which is what has been observed, for example in sickle cell disease. All the evidence available suggests that the code is not overlapping and this is generally accepted among scientists.

...and a degenerate code

The genetic code contains more information than is needed. If you look carefully at **tables A and B**, you will see that often only the first two of the three nucleotides in a codon seem to matter in determining which amino acid results. This may seem a rather useless feature at first, but if each amino acid was produced by only one codon, then any error or mutation could cause havoc. With a degenerate (or redundant) code, if the final base in the triplet is changed, this mutation could still produce the same amino acid and have no effect on the organism. Only methionine and tryptophan are represented by only one codon.

Mutations can happen any time the DNA is copied – the degenerate code at least partly protects living organisms from their effects.

Questions

1　Explain what is meant by the genetic code.

2　What is non-coding DNA?

3　What are the benefits to an organism of having:
  (a) a non-overlapping code?
  (b) a degenerate code?

Key definitions

Translation is the process by which proteins are produced, via RNA, using the genetic code found in the DNA. It takes place on the ribosomes.

Ribosomes are the site of protein synthesis in the cell.

A **triplet code** is the code of three bases, and is the basis of the genetic information in the DNA.

A **gene** is a sequence of bases on a DNA molecule. It contains coding for a sequence of amino acids in a polypeptide chain that affect a characteristic in the phenotype of the organism.

A **codon** is a sequence of three bases in DNA or mRNA.

A **complementary strand** is the strand of RNA formed that complements the DNA acting as the coding strand.

By the end of this section, you should be able to...

- describe the structure of mRNA including nucleotides, the sugar–phosphate backbone and the role of hydrogen bonds

- describe the structure of tRNA including nucleotides, the role of hydrogen bonds and the anticodon

- explain the processes of transcription in the nucleus and translation at the ribosome, including the role of sense and antisense DNA, mRNA, tRNA and the ribosomes

In eukaryotes, the DNA that codes for the individual proteins is in the nucleus of the cell. The ribosomes where proteins are synthesised are in the cytoplasm. DNA from the nucleus has never been detected in the cytoplasm, so the message cannot be carried directly. RNAs (ribonucleic acids) carry the information from the nuclear DNA to the ribosomes.

Different types of RNA

RNA is closely related to DNA (see **Section 1.3.2**). However, it contains a different sugar (ribose) and a different base (uracil instead of thymine). It consists of a single helix and does not form enormous and complex molecules like DNA. The sequence of bases along a strand of RNA relates to the sequence of bases on a small part of the DNA in the nucleus. RNA enables DNA to act as the genetic material. It carries out three main functions in the process of protein synthesis:

- It carries the instructions for a polypeptide from the DNA in the nucleus to the ribosomes where proteins are made.

- It picks up specific amino acids from the protoplasm and carries them to the surface of the ribosomes.

- It makes up the bulk of the ribosomes themselves.

To perform these three very different functions, there are three different types of RNA.

Messenger RNA

Messenger RNA (mRNA) is formed in the nucleus. Whereas a double helix of DNA carries information about a vast array of proteins, a piece of mRNA usually has instructions for one polypeptide. The messenger RNA forms on the template or **antisense strand** of the DNA. The mRNA formed codes for a polypeptide. The coding strand of DNA is known as the **sense** strand of DNA. Any mRNA formed on this strand would be nonsense and would not code for a protein.

Parts of the DNA molecule unravel and are transcribed onto strands of mRNA by an enzyme called **DNA-directed RNA polymerase**. This enzyme is often known as **RNA polymerase**, but the full name tells you it polymerises nucleotide units to form RNA in a sequence determined by the DNA. The complementary bases

in the nucleotides of the DNA and RNA line up alongside each other. RNA nucleotides from the nucleoplasm line up alongside the exposed DNA. Initially hydrogen bonds hold the complementary RNA bases in place. Then DNA-directed RNA polymerase catalyses the formation of phosphodiester bonds between the sugars and phosphate groups of the bases, to form a strand of mRNA. Hydrogen bonds maintain the helical structure of the RNA molecule. Just as in the DNA, the bases of the mRNA form a triplet code and each triplet of bases is a codon. The relatively small mRNA molecules pass easily through the pores in the nuclear membrane, carrying the instructions from the genes in the nucleus to the cytoplasm. They then move to the surface of the ribosomes, where protein synthesis takes place (see **fig A**).

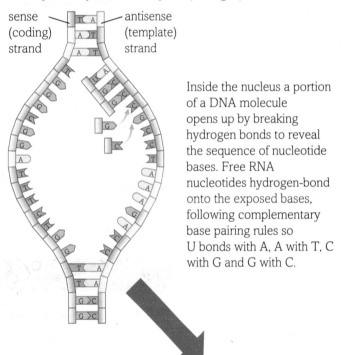

Inside the nucleus a portion of a DNA molecule opens up by breaking hydrogen bonds to reveal the sequence of nucleotide bases. Free RNA nucleotides hydrogen-bond onto the exposed bases, following complementary base pairing rules so U bonds with A, A with T, C with G and G with C.

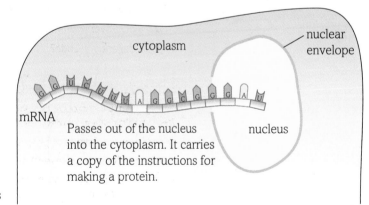

Passes out of the nucleus into the cytoplasm. It carries a copy of the instructions for making a protein.

fig A The transcription of the DNA message. Any mistakes in this process can have fatal consequences for the cell or even the whole organism if the wrong protein is made.

Transfer RNA

Transfer RNA (tRNA) is found in the cytoplasm. It has a complex shape, often described as a clover leaf, that enables it to carry out its function (see **fig B**). This shape is the result of hydrogen bonding between different bases. One part of the tRNA molecule has a sequence of three bases that matches the genetic code of the DNA and corresponds to one particular amino acid. This sequence of three bases is called the **anticodon**. Each tRNA molecule also has a binding site with which it picks up one particular amino acid from the vast numbers always free in the cytoplasm.

The tRNA molecules, each carrying a specific amino acid, line up alongside the mRNA on the surface of the ribosome. The anticodons of the tRNA line up with the codons of the mRNA, held in place by hydrogen bonds between the corresponding bases. Because the anticodon has a sequence of bases that align with the corresponding bases in the mRNA on the ribosomal surface, the correct sequence of amino acids is assembled. Once the amino acids are lined up together, peptide bonds form between them, building up a long chain of amino acids.

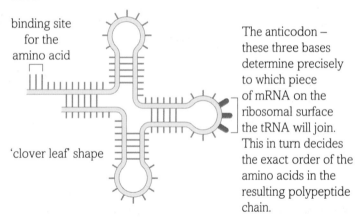

binding site for the amino acid

'clover leaf' shape

The anticodon – these three bases determine precisely to which piece of mRNA on the ribosomal surface the tRNA will join. This in turn decides the exact order of the amino acids in the resulting polypeptide chain.

fig B There are 61 types of tRNA molecules available to carry all the necessary amino acids to the surface of the ribosomes ready for synthesis into protein molecules.

Ribosomal RNA

Ribosomal RNA (rRNA) makes up about 50% of the structure of a ribosome and is the most common form of RNA found in cells. It is made in the nucleus, under the control of the nucleoli, and then moves out into the cytoplasm where it binds with proteins to form ribosomes. The ribosomes consist of a large and a small subunit. They surround and bind to the parts of the mRNA that are being actively translated, and then move along to the next codon. Their job is to hold together the mRNA and tRNA and act as enzymes controlling the process of protein synthesis.

Protein synthesis

To summarise, in the process of protein synthesis the genetic code of the DNA of the nucleus is transcribed onto messenger RNA. This mRNA moves out of the nucleus into the cytoplasm and becomes attached to a ribosome. Molecules of transfer RNA carry individual amino acids to the surface of the ribosome. The tRNA anticodon lines up alongside a complementary codon in the mRNA, held in place by hydrogen bonds while enzymes link the amino acids together. The tRNA then breaks away and returns to the cytoplasm to pick up another amino acid. The ribosome moves along the molecule of mRNA until it reaches the end, leaving a completed polypeptide chain. The message may be read again and again.

Protein synthesis, like many other events in living things, is a continual process. However, it makes it simpler to understand if we look at the two main aspects of it separately. The events in the nucleus involve the transcription of the DNA message (see **fig A**). In the cytoplasm that message is translated into polypeptide molecules and hence into proteins (see **fig C**).

Mass production

The cytoplasm of cells contains many **polysomes**. These are groups of ribosomes joined by a thread of mRNA, and they appear to be a form of mass production of particular proteins. Instead of one ribosome moving steadily along a strand of mRNA and producing its polypeptide and then repeating the process, ribosomes attach in a steady stream to the mRNA and move along one after the other producing lots of identical polypeptides.

This is how the genetic code carried on the DNA is translated into living material by the synthesis of proteins.

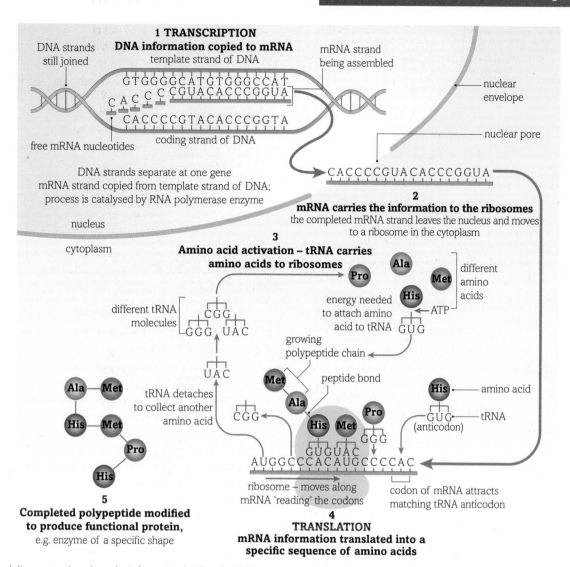

fig C A simplified diagram to show how the information held in the DNA sequence in the nucleus is translated into a sequence of amino acids in a polypeptide chain. In reality, the mRNA strand and the amino acid chain may be thousands of units long.

Questions

1 DNA and RNA are the information molecules of the cell. Explain the differences in the basic structures of these two molecules.

2 In many organisms the DNA is in the nucleus of the cells and the proteins for which it codes are in the cytoplasm. Explain carefully the roles of the following in translating the genetic code into an active enzyme in the cytoplasm of a cell:

(a) DNA

(b) messenger RNA

(c) transfer RNA

(d) ribosomal RNA

Key definitions

Messenger RNA (mRNA) is the RNA formed in the nucleus that carries the genetic code out into the cytoplasm.

The **antisense strand** (template strand) is the DNA strand that codes for proteins.

DNA-directed RNA polymerase (RNA polymerase) is the enzyme that polymerises nucleotide units to form RNA in a sequence determined by the antisense strand of DNA.

Transfer RNA (tRNA) molecules are small units of RNA that pick up particular amino acids from the cytoplasm and transport them to the surface of the ribosome to align with the mRNA.

The **anticodon** is a sequence of three bases on tRNA that are complementary to the bases in the mRNA codon.

Ribosomal RNA (rRNA) is RNA that makes up about 50% of the structure of the ribosome.

Polysomes are groups of ribosomes, joined by a thread of mRNA, that can produce large quantities of a particular protein.

The **sense** strand of DNA has the same base sequence as the mRNA transcribed from the antisense strand.

1.3   6   Gene mutation

By the end of this section, you should be able to...

● explain the term gene mutation and describe base deletions, insertions and substitutions

● explain the effect of point mutations on amino acid sequences as illustrated by sickle cell disease in humans

The genetic code carried on the DNA is translated into living cellular material through protein synthesis. If a single codon is changed or misread during the process, then the amino acid for which it codes may be different. As a result the whole polypeptide chain and indeed the final protein may be altered. A change like this is known as a **mutation**. A mutation is a permanent change in the DNA of an organism. A mutation can happen when the **gametes** (sex cells) form, although they also occur during the division of somatic (body) cells.

A tiny alteration at this molecular level may have no noticeable effect at all – but it may have devastating effects on the whole organism. Many human genetic diseases are the result of random mutations in the genetic material of the gametes, including thalassaemia, in which the blood proteins are not manufactured correctly, or cystic fibrosis, in which a membrane protein does not function properly.

Different types of mutations

Gene mutations involve changes in the bases making up the codons. The chance of a mutation taking place during DNA replication is around 2.5×10^{-8} per base, although estimates vary widely as it is very difficult to measure. Fortunately the body also has its own DNA repair systems. Specific enzymes cut out or repair any parts of the DNA strands that become broken or damaged. In spite of this, some mutations remain and are copied from the DNA when new proteins are made.

Some mutations occur when just one or a small number of nucleotides are miscopied during transcription. These are **point** or **gene mutations**. If you think of the amino acids produced from each codon as the equivalent of the letters of the alphabet, the result of a point mutation is like changing a letter in one word. It may well still make an acceptable word, but the meaning will probably be different. These gene mutations include **substitutions**, where one base substitutes for another, **deletions** where a base is completely lost in the sequence, and **insertions**, when an extra base is added, which may be a repetition of one of the bases already there or a different base entirely.

Chromosomal mutations involve changes in the positions of genes within the chromosomes. This is like rearranging the words within a sentence – if you are lucky they still make sense, but it will not mean the same as the original sentence. Finally there are **whole-chromosome mutations**, where an entire chromosome is either lost during meiosis, which is cell division to form the sex cells, or duplicated in one cell by errors in the process.

This is like the loss or repetition of a whole sentence. For example, Down's syndrome is caused by a whole-chromosome mutation at chromosome 21 – affected individuals have three copies of this chromosome instead of the usual two.

How gene mutations can affect the phenotype

Mutations can be a source of variation within an organism. If the different arrangements of nucleotides code for the same amino acid (see **Section 1.3.5**) a point mutation will have no effect. Very occasionally, a mutation occurs that results in the production of a new and superior protein. This may help the organism gain a reproductive advantage so that it leaves more offspring than other individuals of that species, particularly if environmental conditions change. Most mutations are neutral, meaning that they neither improve nor worsen the chances of survival. Some mutations cause great damage, disrupting the biochemistry of the entire organism. If a base mutation change is in a protein that plays an important role in a cell – for example, the active site of an enzyme – the effect can be catastrophic.

Sickle cell disease – when the code goes wrong

Sickle cell disease is a genetic disease that affects the protein chains making up the haemoglobin in the red blood cells. It is the result of a point mutation. A change of one base in one codon changes a single amino acid in a chain of 147 amino acids – but that change alters the nature of the protein. As a result, the haemoglobin molecules stick together to form rigid rods that give the red blood cells a sickle shape. They do not carry oxygen very efficiently and block the smallest blood vessels. This single tiny change in one nucleotide is enough to cause the people affected severe pain and even death.

Sequence for healthy haemoglobin								
ATG	GTG	CAC	CTG	ACT	CCT	GAG	GAG	TCT
Start	Val	His	Leu	Thr	Pro	Glu	Glu	Ser
Sequence for sickle cell haemoglobin								
ATG	GTG	CAC	CTG	ACT	CCT	GTG	GAG	TCT
Start	Val	His	Leu	Thr	Pro	Val	Glu	Ser

table A The change in the single codon that causes sickle cell disease (the first nine codons only shown).

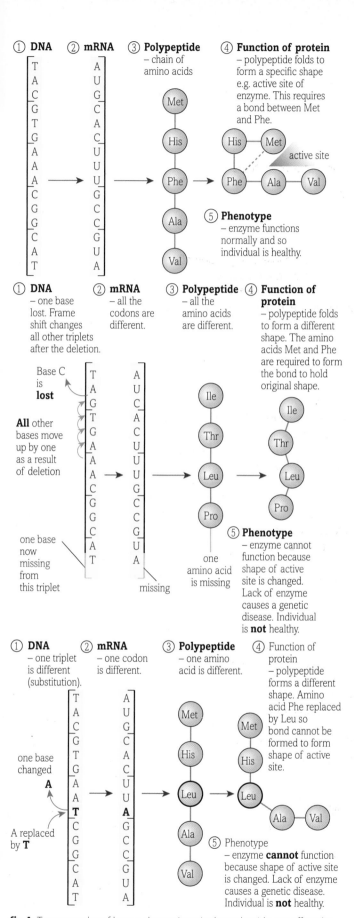

① **DNA** ② **mRNA** ③ **Polypeptide** – chain of amino acids ④ **Function of protein** – polypeptide folds to form a specific shape e.g. active site of enzyme. This requires a bond between Met and Phe.

⑤ **Phenotype** – enzyme functions normally and so individual is healthy.

① **DNA** – one base lost. Frame shift changes all other triplets after the deletion.
② **mRNA** – all the codons are different.
③ **Polypeptide** – all the amino acids are different.
④ **Function of protein** – polypeptide folds to form a different shape. The amino acids Met and Phe are required to form the bond to hold original shape.

Base C is **lost**

All other bases move up by one as a result of deletion

one base now missing from this triplet

missing

⑤ **Phenotype** – enzyme cannot function because shape of active site is changed. Lack of enzyme causes a genetic disease. Individual is **not** healthy.

one amino acid is missing

① **DNA** – one triplet is different (substitution).
② **mRNA** – one codon is different.
③ **Polypeptide** – one amino acid is different.
④ Function of protein – polypeptide forms a different shape. Amino acid Phe replaced by Leu so bond cannot be formed to form shape of active site.

one base changed **A**

A replaced by **T**

⑤ Phenotype – enzyme **cannot** function because shape of active site is changed. Lack of enzyme causes a genetic disease. Individual is **not** healthy.

fig A Two examples of how a change in a single nucleotide can affect the structure and function of the protein formed and hence the phenotype of the individual.

fig B The rigid shape of the red blood cells in sickle cell disease as a result of changes in the haemoglobin molecules prevents them from functioning properly in the body.

Mutations can happen to any cell at any time, though they occur most commonly during the copying of DNA for cell division. Mutations in the body cells can cause problems such as cancer. The most damaging mutations occur in the gametes because they will be passed on to future offspring. These are the mutations that give rise to genetic diseases. Exposure to **mutagens**, such as X-rays, ionising radiation and certain chemicals, increases the rate at which mutations occur. For this reason it is better to keep exposure to these mutagens to a minimum.

Questions

1 Some base mutations will have as big an impact on the way the body works as any chromosomal or whole-chromosome mutation. Others have no effect at all on the organism. Explain.

2 Explain how a change in a single base in the sickle cell mutation has such a dramatic effect on affected individuals.

Key definitions

A **mutation** is a permanent change in the DNA of an organism.

Gametes are haploid sex cells produced by meiosis that fuse to form a new diploid cell (zygote) in sexual reproduction.

A **point mutation (gene mutation)** is a change in one or a small number of nucleotides affecting a single gene.

A **substitution** is a type of point mutation in which one base in a gene is substituted for another.

A **deletion** is a type of point mutation in which a base is completely lost.

An **insertion** is a type of point mutation in which an extra base is added into a gene, which may be a repeat or a different base.

Chromosomal mutations are changes in the position of entire genes within a chromosome.

A **whole-chromosome mutation** is the loss or duplication of a whole chromosome.

Sickle cell disease (sickle cell anaemia) is a human genetic disease affecting the protein chains making up the haemoglobin in the red blood cells.

A **mutagen** is anything that increases the rate of mutation.

1 (a) The diagram below shows the structure of a nucleotide monomer unit from a DNA molecule.

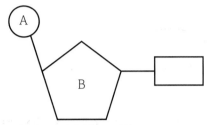

Name the parts labelled A and B. [1]

(b) The table below shows the percentage of different bases present in the DNA from a cow.

Percentage of each base present			
Adenine	Guanine	Thymine	Cytosine
		29	

(i) Complete the table to show the percentage of adenine, guanine and cytosine in the DNA of the cow. [1]
(ii) Explain how you worked out the percentage of guanine present in the DNA of a cow. [3]

[Total 5]

2 The diagram below shows part of a DNA molecule.

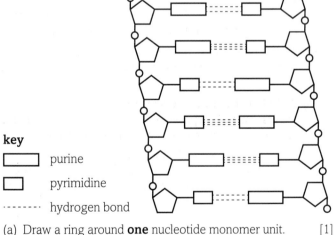

key

☐ purine

☐ pyrimidine

------- hydrogen bond

(a) Draw a ring around **one** nucleotide monomer unit. [1]
(b) Name the two **purine** bases found in DNA. [1]
(c) (i) State where transcription takes place in eukaryotic cells. [1]
(ii) During transcription, part of a DNA molecule unwinds and the DNA strands separate.
Describe the events that follow to produce a messenger RNA (mRNA) molecule. [3]

(d) Oligonucleotides are short chains of nucleotides. Some of these are man-made and have been used as drugs to treat a wide variety of diseases. They work by binding to mRNA or DNA and inhibiting protein synthesis. The drugs are described as antisense drugs when they bind to mRNA and triplex drugs when they bind to DNA.
(i) State which stage of protein synthesis will be inhibited by each of the following
• Antisense drugs
• Triplex drugs [1]
(ii) The table below shows the sequence of bases in part of a molecule of mRNA.
Complete the table to show the sequence of bases in the antisense drug that will bind to this part of the mRNA molecule.

Base sequence on DNA	A	G	U	C	A	U
Base sequence in antisense drug						

[1]
[Total 8]

3 (a) Work by Nachman and Crowell (*Genetics*, September 1, 2000 vol. 156 no. 297–304) using human DNA has estimated an average mutation rate of 2.5×10^{-8} per base. Assuming there are 7×10^9 base pairs in a human diploid cell, calculate the average number of mutations formed each time the cell divides. [2]

(b) If a cell divides 50 times, calculate the total number of mutations accumulated by an average cell. [1]

(c) Explain why most cells still produce proteins that work properly despite this number of mutations. [3]
[Total 6]

4 The diagram below shows some cell structures involved in protein synthesis in eukaryotic cells.

Nucleus   Rough endoplasmic reticulum

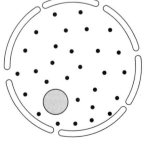
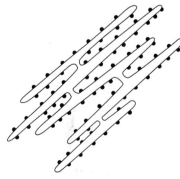

(a) Describe the events that occur inside the nucleus to produce a molecule of messenger RNA (mRNA). [4]

(b) Describe the role of the ribosomes in protein synthesis. [3]

(c) The table below gives some of the base triplets on DNA that code for some amino acids and stop signals.

Base triplet on DNA	Amino acid/stop signal
CCC	Glycine
AAA or AAG	Phenylalanine
AGA or AGC	Serine
GCG	Arginine
TTT	Lysine
ATT or ATC or ACT	Stop signal

The diagram below shows the final base triplets of a gene, labelled T1 to T5, and the complementary messenger RNA (mRNA).

The sequence of amino acids at the end of the protein produced is also shown.

	T1	T2	T3	T4	T5
Last part of the DNA strand:	CCC	GCG	AGC	TTT	

Complementary mRNA: [ ][ ][ ][ ][ ]

Amino acid sequence: [Glycine]—[ ]—[ ]—[ ]

(i) Write in the codons found on the mRNA complementary to the base triplets T1, T2, T3, T4 on the diagram above. [2]

(ii) Using the information in the timetable, complete the amino acid sequence shown in the diagram above. The first one has been done for you. [2]

(iii) Use the information in the table to deduce a base triplet for T5 on the DNA strand. [1]

[Total 12]

6 In DNA, the type of bond that joins deoxyribose sugar to a phosphate group is

(a) a phosphodiester bond

(b) a hydrogen bond

(c) a peptide bond

(d) a glycosidic bond [1]

[Total 1]

7 The sequence **CCGAAACGACTC** on a DNA strand when transcribed would form which mRNA sequence?

(a) CCGUUUCGUCAC

(b) GGCUUUGCUGAG

(c) CCGAAACGACUC

(d) GGCAAAGCAGTG [1]

[Total 1]

8 The strand of DNA that is transcribed is called the

(a) sense strand

(b) antisense strand

(c) same sense strand

(d) missense strand [1]

[Total 1]

9 Meselson and Stahl's experiment showed that DNA replicated semiconservatively. Bacteria grown on a medium containing ^{15}N over many generations would have DNA containing this one isotope of nitrogen. They were then transferred to grow on a medium that contained ^{14}N.

(a) Describe how these enzymes are used in DNA replication
　(i) DNA helicase [2]
　(ii) Polymerase [3]
　(iii) Ligase [2]

(b) Calculate the proportion of bacterial DNA that contains exclusively ^{14}N after the third round of replication. [2]

(c) The strands of DNA were separated according to their density. Explain why it was necessary that ^{15}N is a stable isotope of nitrogen. [1]

(d) If DNA replicated conservatively then the results after the first round of replication would have differed from the observed result with semiconservative replication.

Describe how the DNA strands formed by one round of semiconservative replication differ from those that might have formed from one round of conservative replication. [2]

[Total 12]

TOPIC 1
Biological molecules

Enzymes

Introduction

Around the world, people have observed albino forms in almost every species of vertebrate, including human beings. True albino animals are very striking – they have pure white hair, fur or feathers, usually with pale skin and red or pale blue eyes. Albinism is genetic – it is inherited in the genes passed on from parents to their offspring. But why do albino animals lack colour? It is all down to the lack of a single enzyme – tyrosinase. This enzyme is a key factor in the production of the pigment melanin (and other pigments) from the amino acid tyrosine. Differing amounts of melanin, combined with other pigments, result in a wide range of hair, skin and feather colours. If the enzyme is lacking, the animal lacks pigment and is albino.

In this chapter you will be looking at enzymes – their structure, how they work and what happens when they are inhibited. You will look at the varying roles of enzymes in the body and how they are named. You will go on to discover how enzymes work and how their mechanism of action is related to the shape of the active site produced within the tertiary structure of the protein itself. By looking at the evidence you will see how our models of enzyme action have changed from the relatively simplistic lock-and-key theory to the more complex induced-fit model.

Measuring the rate of enzyme controlled reactions is key to understanding the factors that affect them. You will be considering the practical difficulties of doing this and how they can be overcome. Looking at the factors that affect the rate of an enzyme controlled reaction helps to build up our model of how enzymes act as catalysts in biological systems.

Understanding how enzyme action can be inhibited by other molecules is another way of developing an understanding of how enzymes work. You will consider competitive, non-competitive and irreversible inhibition, as well as considering the situation when the end products of a long chain of reactions inhibit an enzyme earlier in the process.

All the maths you need

- Recognise and make use of appropriate units in calculations (*e.g. the units for the rate of reaction of an enzyme*)
- Use of percentages (*e.g. calculating percentage yields in different enzyme controlled reactions*)
- Use of appropriate number of significant figures (*e.g. understand that results for enzyme rate experiments can be reported only to the limits of the least accurate measurement*)
- Find arithmetic means (*e.g. the mean of a range of data when investigations are repeated*)
- Plot a range of data in an appropriate format (*e.g. enzyme activity over time represented on a graph*)
- Solve algebraic equations (*e.g. calculate the rate of enzyme reactions*)
- Plot two variables from experimental or other data (*e.g. select an appropriate format for presenting data from experimental investigations into enzyme controlled reactions*)
- Understand that $y = mx + c$ represents a linear relationship and predict or sketch the shape of a graph with a linear relationship (*e.g. the effect of substrate concentration on the rate of an enzyme controlled reaction with excess enzyme*)
- Calculate the rate of change from a graph showing a linear relationship (*e.g. the rate of an enzyme controlled reaction*)
- Draw and use the slope of a tangent to a curve as a measure of rate of change and use this method to measure the gradient at a point on a curve (*e.g. amount of product formed plotted against time when the concentration of enzyme is fixed*)

What will I study later?

- The importance of enzymes in controlling reactions inside and outside of cells
- The position of enzymes within cell organelles
- Enzymes in lysosomes and apoptosis
- The role of enzymes in fertilisation of the female gamete in mammals and in plants
- The importance of ATPase in the hydrolysis of ATP to provide accessible energy for biological processes
- The role of enzymes in nervous transmission and neuromuscular junctions
- Enzymes in the blood clotting cascade
- The importance of enzymes in cellular respiration (A level)
- The importance of enzymes in photosynthesis (A level)
- The role of enzymes such as DNA ligase and restriction endonucleases in gene technology (A level)
- The importance of enzymes in hormonal responses, e.g. adrenaline (A level)

What have I studied before?

- That enzymes are proteins
- The basic mechanisms of enzyme action, including the active site
- That enzymes have specificity
- Factors affecting the rate of enzymatic reactions

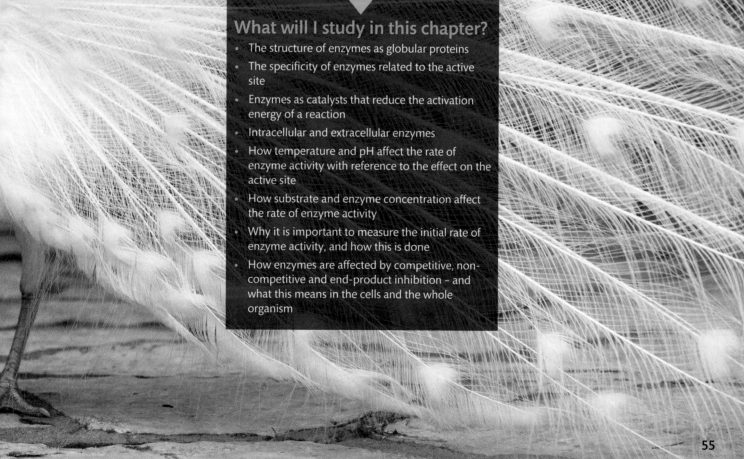

What will I study in this chapter?

- The structure of enzymes as globular proteins
- The specificity of enzymes related to the active site
- Enzymes as catalysts that reduce the activation energy of a reaction
- Intracellular and extracellular enzymes
- How temperature and pH affect the rate of enzyme activity with reference to the effect on the active site
- How substrate and enzyme concentration affect the rate of enzyme activity
- Why it is important to measure the initial rate of enzyme activity, and how this is done
- How enzymes are affected by competitive, non-competitive and end-product inhibition – and what this means in the cells and the whole organism

By the end of this section, you should be able to...

● describe the structure of enzymes as globular proteins

● explain the concept of specificity

● recognise that enzymes catalyse a wide range of intracellular reactions as well as extracellular ones

What is an enzyme?

A **catalyst** is a substance that changes the rate of a reaction without changing the substances produced. The catalyst is unaffected at the end of the reaction and can be used again. **Enzymes** are biological catalysts, which control the rate of the reactions that take place in individual cells and in whole organisms. Under the conditions of temperature and pH found in living cells, most of the reactions that provide cells with energy and produce new biological material would take place very slowly – too slowly for life to exist. Enzymes make life possible by speeding up the chemical reactions in cells without changing the conditions in the cytoplasm.

Enzymes are globular proteins (see **Section 1.2.4**), produced during protein synthesis as the mRNA transcribed from the DNA molecule is translated (see **Section 1.3.5**). They have a very specific shape as a result of their primary, secondary, tertiary and quaternary structures (see **Section 1.2.4**), and this means each enzyme will only catalyse a specific reaction or group of reactions. We say enzymes show great **specificity**. Changes in temperature and pH affect the efficiency of an enzyme because they affect the intramolecular bonds within the protein that are responsible for the shape of the molecule.

Within any cell many chemical reactions are going on at the same time. Those reactions that build up new chemicals are known as **anabolic reactions** ('ana' means up, as in 'build up'). Those that break substances down are **catabolic reactions** ('cata' means down). The combination of these two processes results in the complex array of biochemistry that we refer to as **metabolism**. Most of the reactions of metabolism occur not as single events but as part of a sequence of reactions known as a **metabolic chain** or **metabolic pathway**. We usually think of enzymes speeding up reactions but sometimes they act to slow them down, or stop them completely.

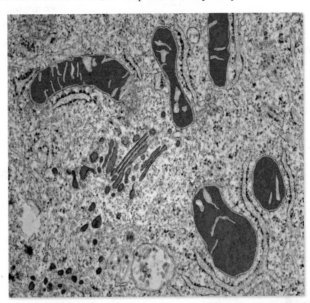

fig A Each cell contains several hundred different enzymes to control the multitude of reactions going on inside.

Naming enzymes

In the study of biology, in medicine, in cellular and genetic research and in industries that use biotechnology, it is important to be able to refer to the action of specific enzymes. To do this we need to understand how enzymes are named.

Many of the enzymes found in animals and plants work inside the cells. These are known as **intracellular enzymes**, for example DNA polymerase and DNA ligase. Cells secrete other enzymes that have an effect beyond the boundaries of the cell membrane. These are **extracellular enzymes**. The digestive enzymes and lysozyme, the enzyme in your tears, are well-known examples of these.

Most enzymes – both intracellular and extracellular – have several names including:

- a relatively short recommended name, which is often the name of the molecule that the enzyme works on (the substrate) with '-ase' on the end, or the substrate with an indication of what it does, e.g. creatine kinase

- a longer systematic name describing the type of reaction being catalysed, e.g. ATP:creatine phosphotransferase

- a classification number, e.g. EC 2.7.3.2.

Some enzymes, such as urease, ribonuclease and lipase, are known by their recommended names. But there are still some enzymes that are known by common but uninformative names – trypsin and pepsin for example. However, the names of most enzymes give you useful information about the role of the enzyme in the cell or the body.

Did you know?

The discovery of enzymes

In 1835 people noticed that starch is broken down to sugars more effectively by malt (sprouting barley) than by sulfuric acid.

People also suspected there were 'ferments' in yeast (a single-celled fungus) that turned sugar to alcohol and in 1877 the name enzyme (literally 'in yeast') was introduced. In 1897 Eduard Buchner (1860–1917) extracted a 'juice' from yeast cells that would breakdown various sugars outside a living cell.

In 1926 James B. Sumner (1887–1955) extracted the first pure, crystalline enzyme from jack beans. It was urease, the enzyme that catalyses the breakdown of urea. Sumner found the crystals were protein and concluded that enzymes must therefore be proteins. Unfortunately no-one believed the young researcher at the time, because many established scientists had been trying and failing to isolate enzymes for years. However, 20 years later Sumner received a Nobel Prize for his ground-breaking work.

fig B Pure urease does not look very exciting, but the ability to isolate and extract enzymes has revolutionised our understanding of biology and the way we can use enzymes in industry.

Questions

1 From which organisms were the first enzymes isolated?

2 What is the difference between an intracellular enzyme and an extracellular enzyme?

3 Investigate Sumner's work and discover which scientists were particularly against his ideas and why.

Key definitions

A **catalyst** is a substance that speeds up a reaction without changing the substances produced or being changed itself.

Enzymes are proteins that have a very specific shape as a result of their primary, secondary, tertiary and quaternary structures. They act as biological catalysts and each enzyme will only catalyse a specific reaction or group of reactions.

Specificity is the characteristic of enzymes that means that, as a result of the very specific shapes resulting from their tertiary and quaternary structures, each enzyme will only catalyse a specific reaction or group of reactions.

An **anabolic reaction** is the reaction that builds up (synthesises) new molecules in a cell.

A **catabolic reaction** is a reaction which breaks down substances within a cell.

Metabolism is the sum of the anabolic and catabolic processes in a cell.

A **metabolic chain (metabolic pathway)** is a series of linked reactions in the metabolism of a cell.

Intracellular enzymes are enzymes that catalyse reactions within the cell.

Extracellular enzymes are enzymes that catalyse reactions outside of the cell in which they were made.

By the end of this section, you should be able to...

● explain how enzymes act as catalysts by reducing the activation energy of reactions

● explain how the initial rate of enzyme activity can be measured and why this is important

● explain how different factors affect the rate of enzyme activity

For a chemical reaction to take place, the reacting molecules must have enough energy to break the chemical bonds that hold them together. A simple model is that the reaction has to get over an 'energy hill', known as the **activation energy**, before it can get started.

Raising the temperature increases the rate of a chemical reaction by giving more molecules sufficient energy to react. However, living cells could not survive the temperatures needed to make many cellular reactions fast enough – and the energy demands to produce the heat would be enormous. Enzymes solve the problem by lowering the activation energy needed for a reaction to take place (see **fig A**).

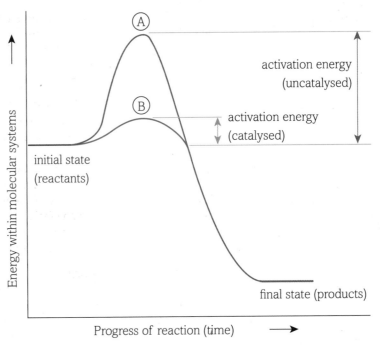

(A) = Energy of transition state in uncatalysed reaction.

(B) = Energy of transition state, i.e. enzyme/substrate complex, during catalysed reaction.

fig A Energy diagram to show the difference between an uncatalysed and a catalysed reaction.

How do enzymes work?

To lower the activation energy and catalyse a reaction, enzymes form a complex with the **substrate** or substrates of the reaction. A simple picture of enzyme action in a catabolic reaction is:

$$\text{substrate} + \text{enzyme} \rightleftharpoons \text{enzyme/substrate complex} \rightleftharpoons \text{enzyme} + \text{products}$$

Once the products of the reaction are formed they are released and the enzyme is free to form a new complex with more substrate. How does this relate to the structure of the enzyme? The **'lock-and-key hypothesis'** gives us a simple model that helps us understand what happens (see **fig B**). Within the globular protein structure of each enzyme is an area known as the **active site** that has a very specific shape. Only one substrate or type of substrate will fit the shape of the gap, and it is this that gives each enzyme its specificity. Just as a key fits into a lock, so the enzyme and substrate slot together to form a complex.

The formation of the enzyme/substrate complex lowers the activation energy of the reaction. The active site affects the bonds in the substrate, making it easier for them to break, and the reacting substances are brought close together, making it easier for bonds to form between them. Once the reaction is complete the products are no longer the right shape to stay in the active site and the complex breaks up, releasing the products and freeing the enzyme for further catalytic action.

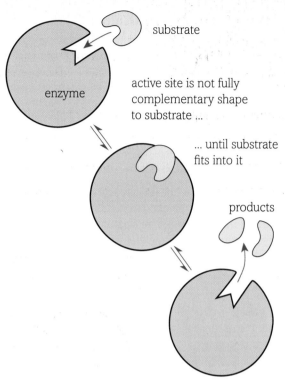

fig C The induced-fit theory of enzyme action proposes that the catalytic groups of the active site are not brought into their most active positions until a substrate is bound to the site, inducing a change in shape.

Measuring reaction rate

When scientists are investigating enzymes and how they act as catalysts, they frequently measure the reaction rate. For example, one practical way of demonstrating the effect of an enzyme on a reaction is to measure the rate of the reaction with and without the enzyme. Using this method it has been shown that when urea breakdown is catalysed by urease extracted from the jack bean, the rate of the reaction increases by a factor of 10^{14}. Enzymes are such efficient catalysts that they generally increase reaction rates by factors from 10^8 to 10^{26}. This is why only tiny amounts of most enzymes are needed.

Much of the evidence for the structure of enzymes and the way this relates to their functions comes from practical investigations into the effect of different factors on the rate of enzyme-catalysed reactions. To investigate the way a factor affects the rate of reaction, biologists measure the **initial rate of reaction** each time the independent variable is changed. Every other factor must be kept the same so that any changes are the result of changing the one variable.

It is important to provide a large excess of substrate in enzyme experiments, unless the effect of substrate concentration is under investigation. The initial rate of reaction is when the reaction proceeds at its fastest rate. This gives the maximum reaction rate for an enzyme under particular conditions, for example changing pH, temperature or substrate concentration.

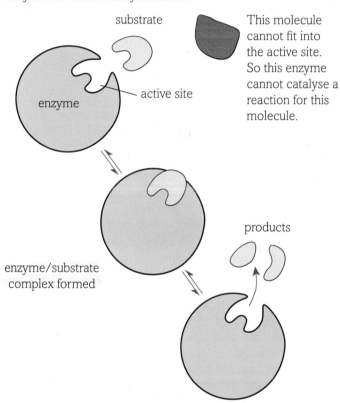

fig B The lock-and-key hypothesis underpins our understanding of how enzymes work.

The lock-and-key hypothesis fits most of our evidence about enzyme characteristics. However, it is now thought to be an over-simplification. Evidence from X-ray crystallography, chemical analysis of active sites and other techniques suggests that the active site of an enzyme is not simply a rigid shape. In the **induced-fit hypothesis**, generally accepted as the best current model of enzyme action, the active site still has a distinctive shape and arrangement, but it is a flexible one. Once the substrate enters the active site, the shape of the site is modified around it to form the active complex. Once the products have left the complex the enzyme reverts to its inactive, relaxed form until another substrate molecule binds (see **fig C**).

What do we know about enzymes?

Our current model of enzymes is that they are globular proteins (see **Section 1.2.4**), which contain an active site that is vital to the functioning of the enzyme. The active site is a small depression on the surface of the molecule that has a specific shape because of the way the whole large molecule is folded. Anything affecting the shape of the protein molecule affects its ability to do its job, which indicates that the three-dimensional (3D) nature of the molecule is important to the way it works. A change in shape changes the shape of the active site as well – and so the enzyme can no longer function.

Enzymes change only the rate of a reaction. They do not change or contribute to the end products that form, or affect the equilibrium of the reaction. They act purely as catalysts and not as modifying influences in any other way.

Evidence for the relationship between the structure and functions of enzymes

Observing the factors that affect the rate of enzyme activity gives an insight into the relationship between the structure of an enzyme and the way it functions.

- Enzymes speed up reactions to such an extent that only minute amounts of them are needed to catalyse the reaction of many substrate molecules into products. This is described by the **molecular activity** or **turnover number** of an enzyme, which measures the number of substrate molecules transformed per minute by a single enzyme molecule. The number of molecules of hydrogen peroxide catalysed by the enzyme catalase extracted from liver cells has been reported as 6×10^6 in 1 minute. Most enzymes would catalyse thousands of molecules per minute rather than millions. If every enzyme molecule is involved in a reaction, it will not go any faster unless there is an increase in the enzyme concentration. In other words, enzyme controlled reactions are affected by the concentration of the enzyme.
- Enzymes are very specific to the reaction that they catalyse. Inorganic catalysts such as platinum frequently catalyse many different reactions, often only at extremes of temperature and pressure. In comparison, some enzymes are so specific that they will catalyse only one particular reaction. Others are specific to a particular group of molecules that are all of similar shape, or to a type of reaction that always involves the same groups. This suggests that there is a physical site within the enzyme with a particular shape into which a specific substrate will fit.
- The number of substrate molecules present (the concentration of the substrate) affects the rate of an enzyme-catalysed reaction. Take a simple reaction where substrate A is converted to product Z. If the concentration of A increases, the rate of the enzyme-catalysed reaction A → Z increases – but only for so long. Then the enzyme becomes saturated – all of the active sites are occupied by substrate molecules – and a further increase in substrate concentration will not increase the rate of the reaction further (see **fig D**). At this point only an increase in enzyme concentration will increase the rate of the reaction.

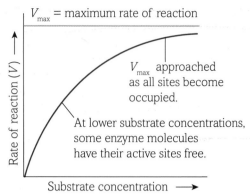

fig D The effect of substrate concentration on an enzyme-catalysed reaction, showing how the enzyme becomes saturated with substrate molecules.

- Temperature affects the rate of an enzyme-catalysed reaction in a characteristic way. Temperature affects all reactions because the number of successful collisions leading to a reaction increases at higher temperatures. The effect of temperature on the rate of any reaction can be expressed as the **temperature coefficient, Q_{10}**. This is expressed as:

$$Q_{10} = \frac{\text{rate of reaction at } (x + 10)\,°C}{\text{rate of reaction at } x\,°C}$$

Between about 0 °C and 40 °C, Q_{10} for any reaction is 2 – the rate of the reaction doubles for every 10 °C rise in temperature. However, outside this range, Q_{10} for enzyme-catalysed reactions in human beings decreases markedly, whilst Q_{10} for other reactions changes only slowly. The rate of enzyme-catalysed reactions in human beings falls as the temperature rises, and at about 60 °C the reaction stops completely in most cases. At temperatures over 40 °C most proteins, including most enzymes, start to lose their tertiary and quaternary structures – they denature. When enzymes denature, the shape of the active site changes and so they lose their ability to catalyse reactions. There are some exceptions to this rule. For example, the enzymes of thermophilic bacteria, which live in hot springs at temperatures of up to 85 °C, are able to work at very high temperatures. They are made of temperature-resistant proteins that contain a very high density of hydrogen bonds and disulfide bonds, which hold them together even at high temperatures (see **Section 1.2.4**). However, the optimum temperature of the enzymes of many organisms, including cold water fish and many plants, is much lower than 40 °C.

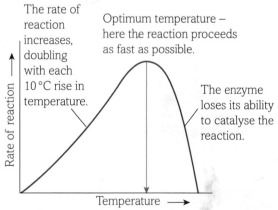

fig E The effect of temperature on the rate of a typical enzyme-catalysed reaction. All other factors must be kept constant.

pH also has a major effect on enzyme activity by affecting the shape of protein molecules. Different enzymes work in different ranges of pH, because changes in pH affect the interactions between R groups, for example hydrogen bonds and ionic bonds that hold the 3D structure of the protein together. The optimum pH for an enzyme is not always the same as the pH of its normal surroundings. This seems to be one way in which cells control the effects of their intracellular enzymes, increasing or decreasing their activity by minute changes in the pH.

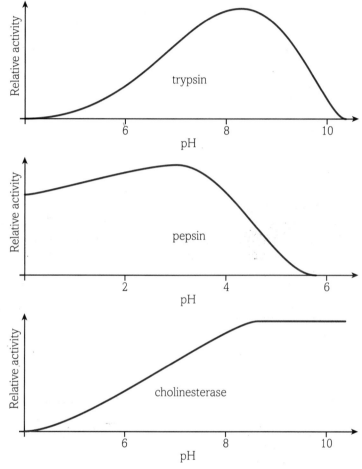

fig F Different enzymes work best at different pH levels. All other factors must be kept constant.

Did you know?

RuBisCo – a key but inefficient enzyme for life

Ribulose bisphosphate carboxylase/oxygenase (known as RuBisCo) is vitally important in photosynthesis. It is the enzyme that catalyses the fixing of carbon dioxide from the air into the biochemistry of sugar formation. Without this enzyme, life as we know it would not exist. But RuBisCo is a remarkably inefficient enzyme.

- Most enzymes catalyse about 1000 reactions per second, RuBisCo only catalyses about 3. Plant cells overcome this by making very large quantities of RuBisCo – about half of the protein in a photosynthetic plant cell is this enzyme.
- The active site of most enzymes is very specific. RuBisCo binds to carbon dioxide molecules in photosynthesis but it can also bind to oxygen molecules in a process called photorespiration. Its affinity for carbon dioxide is about 80 times greater than for oxygen – but there is much more oxygen available so about 25% of RuBisCo binds to oxygen.

Scientists believe RuBisCo evolved in an atmosphere containing very little oxygen and much more carbon dioxide than it does today so oxygen-binding was not a disadvantage at the time and so it was not selected against in evolution.

Questions

1 (a) Summarise the characteristics of enzymes.
  (b) Explain how each characteristic of enzymes provides evidence for the induced-fit hypothesis.

2 Plan a practical investigation into the effect of temperature on enzyme activity.

Key definitions

Activation energy is the energy needed for a reaction to get started.

A **substrate** is the molecule or molecules on which an enzyme acts.

The **lock-and-key hypothesis** is the model that explains enzyme action by an active site in the protein structure that has a very specific shape. The enzyme and substrate slot together to form a complex as a key fits in a lock.

An **active site** is the area of an enzyme that has a specific shape into which the substrate(s) of a reaction fit.

The **induced-fit hypothesis** is a modified version of the lock-and-key hypothesis for enzyme action where the active site is considered to have a more flexible shape. Once the substrate enters the active site, the shape of that site is modified around it to form the active complex. Once the products have left the complex, the enzyme reverts to its inactive, relaxed form.

The **initial rate of reaction** is the measure taken to compare the rates of enzyme controlled reactions under different conditions.

Molecular activity (turnover number) is the number of substrate molecules transformed per minute by a single enzyme molecule.

The **temperature coefficient (Q_{10})** is the measure of the effect of temperature on the rate of a reaction.

By the end of this section, you should be able to...

● describe how enzymes can be affected by competitive, non-competitive and end-product inhibition

We can learn more about enzymes and how they work by looking at evidence from substances that stop the enzymes from working. These are called **enzyme inhibitors**. When we look at how inhibitor molecules interfere with the catalytic powers of an enzyme, we can get more evidence about the way they carry out their functions. There are two main types of inhibition, **reversible inhibition** and **irreversible inhibition**.

Reversible inhibition of enzymes

When an inhibitor affects an enzyme in a way that does not permanently damage it, this is reversible inhibition. When a reversible inhibitor is removed, the enzyme can function normally again. Reversible inhibition is a common feature of metabolic pathways, and it provides a key way of controlling reactions, as you will see. There are two major forms of reversible inhibition – **competitive inhibition** and **non-competitive inhibition**.

Competitive inhibition

In competitive, reversible inhibition, the inhibitor molecule is similar in shape to the substrate molecule. It competes with the substrate for binding at the active sites of the enzymes, forming an enzyme/inhibitor complex. If the amount of inhibitor is fixed, the percentage of inhibition can be reduced by increasing the substrate concentration. The two molecule types are competing for the same active site. The more substrate molecules there are, the less likely it is that inhibitor molecules will bind to the active site.

Non-competitive inhibition

In non-competitive reversible inhibition, the inhibitor may form a complex with either the enzyme itself or with the enzyme/substrate complex. This shows that the inhibitor is not competing for the active site. It joins to the enzyme molecule elsewhere. This is confirmed by the fact that only the concentration of inhibitor affects the level of inhibition. The concentration of the substrate makes no difference to how much inhibition occurs. The best model for how this inhibition works is that the presence of the inhibitor on the enzyme or enzyme/substrate complex deforms or changes the shape of the active site so that it can no longer catalyse the reaction. **Fig A** shows the differences between competitive and non-competitive inhibition.

competitive inhibition

non-competitive inhibition

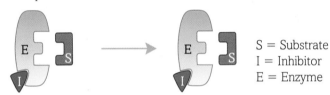

S = Substrate
I = Inhibitor
E = Enzyme

fig A Competitive inhibitors bind at the active site, non-competitive inhibitors do not.

Irreversible inhibition of enzymes

In irreversible inhibition the inhibitor combines with the enzyme by permanent covalent bonding to one of the groups vital for catalysis to occur. It changes the shape and structure of the molecule in such a way that it cannot be reversed – the enzyme is inactivated permanently. Irreversible inhibition tends to occur more slowly than the other forms of inhibition, but its effects are much more devastating and are never used within the cells to control metabolism.

Arsenic, cyanide and mercury are poisonous because they exert irreversible inhibition on enzyme systems. Some of the nerve gases used in chemical warfare also work in this way. They combine with and completely inactivate enzymes such as acetyl cholinesterase that break down chemicals used to transfer impulses from the nervous system to the muscles of the body. The normal function of acetylcholinesterase is to destroy the neurotransmitter called acetylcholine at the junctions between neurones and muscle cells. It does this as soon as an impulse has been passed from a nerve to a muscle. When the enzyme is inhibited the impulse continues. The muscles go into prolonged spasms causing death because breathing and swallowing become impossible.

End-product inhibition and the regulation of the cell

As you know, hundreds of chemical reactions are going on within a cell at any one time, their rate controlled by the action of enzymes. A similar number of reactions occurring in a very small space in a laboratory would, without doubt, end in total chaos if not a large explosion. So how do cells manage their reactions in such a controlled way? There are many factors involved. Membrane compartments keep reactions apart. Variations in pH can change the rate of enzyme-catalysed reactions, and the amount of substrate available is another mechanism at work. But one of the most important methods of control is that exerted by the regulatory enzymes.

Regulatory enzymes often have a site, seperate from the active site, to which another molecule can bind and bring about non-competitive inhibition. They are widely found in complex metabolic pathways such as photosynthesis and respiration.

In **end-product inhibition** the regulatory enzyme is found near the beginning of the pathway. It is inhibited by one of the end products of the chain. There are some very important examples of end-product inhibition in the pathways of cellular respiration in all organisms. Phosphofructokinase (PFK) is an enzyme involved in the production of ATP in the process of glycolysis in cellular respiration (see **Book 2, Section 5.1.2**). PFK controls the rate of respiration by end-product inhibition. It is inhibited by ATP, which binds non-competitively and changes the shape of the active site. If the ATP concentration goes up, PFK is inhibited and cellular respiration slows down. As ATP levels fall, ATP molecules detach from PFK and the enzyme becomes active again. Rates of celluar respiration – and so ATP production – increase.

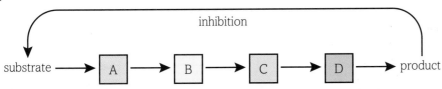

fig B Feedback control gives a simple and effective way of controlling the rate of several reactions at once.

Questions

1 What is the difference between reversible and irreversible enzyme inhibition?

2 What are the main differences between reversible competitive and non-competitive inhibition in enzymes and how does this affect the control of reactions within a cell?

Key definitions

Enzyme inhibitors are substances that slow down enzymes or stop them from working.

Reversible inhibition is inhibition of the action of an enzyme by an inhibitor that does not permanently affect the functioning of the enzyme and can be removed from the enzyme. It is often used to control reaction rates within a cell.

Irreversible inhibition is inhibition of the action of an enzyme that is permanent and cannot be undone. It is never used within cells to control the rate of reactions.

Competitive inhibition is inhibition in which the inhibitor molecule is similar in shape to the substrate molecule and competes with it for the active site of the enzyme (affected by both inhibitor and substrate concentrations).

Non-competitive inhibition is inhibition in which the inhibitor does not compete for the active site but forms a complex with the enzyme or enzyme/substrate complex and changes the shape of the active site so it can no longer catalyse the reaction (affected only by concentration of inhibitor).

Regulatory enzymes are enzymes that have a site separate to the active site where another molecule can bind to have either an activating or inhibitory effect.

End-product inhibition is a control system in many metabolic pathways in which an enzyme at the beginning of the pathway is inhibited by one of the end products of the reaction.

THINKING BIGGER

RAW ENZYMES – REALLY?

The enzymes made by the cells of your body are vitally important. Inside your cells, they control all the reactions of life. Outside your cells they are particularly important in the digestion of your food. The internet is a great source of information but not all of it is reliable. Read the following texts about food and enzymes, based on a number of different websites promoting 'good health'…

Site 1

Each person is born with a limited enzyme-producing capacity. Your life expectancy depends on how well you preserve this enzyme potential. You need to take in enzymes from the food you eat. If you don't take in enough enzymes, it imposes a great strain on your digestive system because it has to produce all the enzymes you need. This in turn reduces the numbers of enzymes available for the metabolic reactions taking place in your cells – and this is the root cause of most chronic health problems. The solution is simple: eat at least 75% of your food raw to make use of the enzymes in the food, eat less, chew your food well and don't chew gum!

Site 2

When food is cooked, enzymes are destroyed by the heat. Enzymes help us digest our food. Enzymes are proteins, and they work because they have a very specific 3D structure in space. Once they are heated much above 118 degrees, this structure can be changed so they no longer work. Cooked foods contribute to chronic illness, because their enzyme content is damaged and so we have to make our own enzymes to process the food. This uses up valuable metabolic enzymes. It takes a lot more energy to digest cooked food than raw food – the evidence being that raw food passes through the digestive tract about 50% faster than cooked food. Eating enzyme-dead (cooked!) foods overworks and eventually exhausts your pancreas and other organs. Many people progressively lose the ability to digest their food after years of eating cooked and processed food.

Site 3

Enzymes are an essential part of a healthy diet. As an expert explains, 'Science cannot duplicate enzymes. Only raw food has functional living enzymes. The chain reaction generated by enzymes helps to send fats to where they are needed in our body, instead of being stored'.

fig A The cells of raw fruit and vegetables are full of enzymes – but how much use are they to you?

Where else will I encounter these themes?

Let us start by considering the nature of the writing in these articles:

1. The information given above comes from a number of different web sites that promote 'healthy living'. Think about the way they are using scientific information as you try and answer the following questions:

   a. Who do you think these web resources are aimed at?

   b. Do you think that the people producing these resources are writing objectively? Explain your answer.

   c. What tactics are used to try to persuade people that eating raw food provides you with useful enzymes and that cooking food is bad for you?

> **Command word**
> If the word explain is used in a question your answer should clearly describe the thing you are trying to explain. You'll need to use your reasoning and maybe examples to support your point.

Now let us have a look at the biology. Your knowledge of biochemistry is now at a level that allows you to read this article with a scientific mind!

2. Make a table to identify the biologically correct information and the biologically suspect information in the articles.

3. Do you think the people writing these web resources are real biologists or doctors? Explain your opinion.

4. Write a blog post describing the dangers of articles like these and putting right all of the biological misconceptions you found in question 2.

Activity

Enzymes are vital for life. A healthy diet provides your body with the materials it needs to make enzymes – but you do not directly use the enzymes in the food that you eat.

Prepare a three minute talk for a debate titled 'Raw food – the only healthy way to support your enzymes.'

Choose whether you want to support this idea or oppose it.

Focus on the biology of enzymes and of the compounds that make up your food. Whichever side you choose your argument must be backed up by good scientific evidence.

> Consider what you have learned about enzymes and their roles in the cells and in the digestive systems of organisms, including people. You can also do research to find out more, but make sure that your sources are reliable!

● Based on a number of different websites promoting 'good health'.

1 Amylase is an enzyme that breaks down starch to maltose. A student carried out an investigation to determine the effect of copper ions on the activity of this enzyme. She added different concentrations of copper ions and timed how long it took the amylase to break down starch.

The results of this investigation are shown in the graph below.

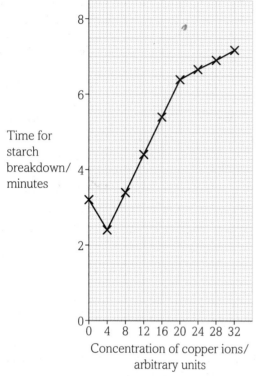

(a) Describe a test that could be used to show that starch has been broken down. [3]

(b) Describe the effect that an increase in the concentration of copper ions has on the **activity** of amylase. [3]

(c) The student suggested that the copper ions were acting as an active site-directed inhibitor at concentrations above 4 arbitrary units. Explain what is meant by the term **active site-directed inhibition**. [3]

(d) The student then investigated the initial rate of reaction using amylase and different concentrations of starch. She did this first with copper ions present and then with no copper ions present. The results are shown in the graph below.

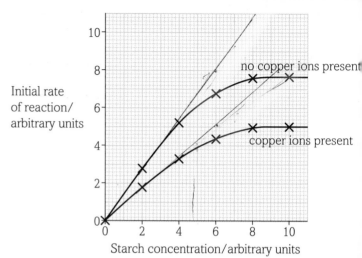

(i) Explain why the **initial** rate of reaction was measured in this investigation. [2]

(ii) State why the results do **not** support the hypothesis that copper ions are an active site-directed inhibitor of amylase. [1]

[Total: 12]

2 (a) The graph below shows the change in energy that takes place during a chemical reaction.

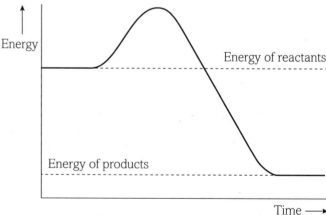

(i) With reference to enzyme activity, explain the meaning of each of the following terms.
 • Activation energy
 • Catalyst [4]

(ii) On the graph above, draw the energy changes that would take place if the same chemical reaction was catalysed by an enzyme. [2]

(b) An experiment was carried out to determine the effect of temperature on the activity of a protein-digesting enzyme (a protease). Solutions of the protease were incubated with a protein called gelatine at three temperatures: 20 °C, 30 °C, and 40 °C. The concentration of amino acids were measured over a 48-hour period. The results are shown in the graph below.

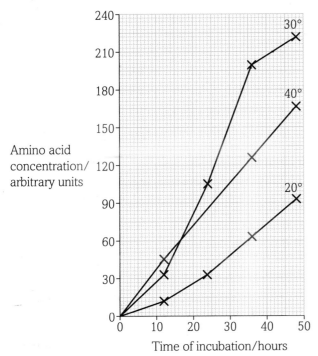

(i) Name the type of reaction catalysed by this protease. [1]
(ii) Name the bond broken by the protease. [1]
(iii) Calculate the mean rate of production of amino acids at 40 °C during the first 36 hours of incubation. Show your working and give your answer in arbitrary units hr^{-1}. [2]
(iv) The optimum temperature for this reaction is 30 °C. Explain the shape of the curve at this temperature. [2]

[Total: 12]

3 Trypsin is a protease enzyme that catalyses the breakdown of proteins. An investigation was carried out to study the effect of pH on the activity of trypsin. The source of protein in this investigation was milk powder mixed in distilled water. This gives a white, cloudy suspension.

A 2% solution of trypsin was prepared and placed in a waterbath at 45 °C. A 10% suspension of milk powder was prepared separately and 3 cm^3 samples of this suspension were mixed with 3 cm^3 of a pH buffer solution and placed in a waterbath at 45 °C. Buffer solutions between pH 5 and pH 9 were used.

When all the mixtures had reached a temperature of 45 °C, 0.5 cm^3 of the trypsin solution was added to the suspension and the time taken for the suspension to clear was recorded.

The results of this experiment are shown in the table below.

pH of suspension	Time taken for suspension to clear/min
5	10.82
6	6.68
7	1.38
8	0.71
9	1.42
10	7.80

(a) (i) Explain how an enzyme, such as trypsin, digests (breaks down) the protein in the milk powder. [3]
(ii) Describe why the cloudy suspension goes clear when mixed with the enzyme. [1]
(b) (i) Use the information in the table to describe the effect of pH on the activity of trypsin. [2]
(ii) Explain how pH affects the activity of enzymes. [3]
(c) Explain why the experiment was carried out in a water bath at 45 °C. [2]

[Total: 11]

TOPIC 2
Cells and viruses

Eukaryotic cells

Introduction

In October 1951, Henrietta Lacks, a young, 31-year-old American black woman, died of an aggressive cancer of the cervix. Before she died, doctors took a sample of her cells, which became the first human cells to grow successfully in culture. Known as HeLa cells (for Henrietta Lacks), her cells are still alive and well. Ever since they were first cultured, HeLa cells have played an important part in cell biology and medical research. The cells have been infected with viruses, and played a big role in the development of the first polio vaccine. HeLa cells reproduce themselves quickly and reliably and they have travelled all over the world. They have been used to investigate radiation damage and they have even gone up into space.

In this chapter you will be discovering more about eukaryotic cells like Henrietta's and the sub-cellular membrane-bound organelles they contain. The animals, plants and fungi in the world around you are all eukaryotic. Without microscopes we could not even see most cells, and so you will learn something of how both light and electron microscopes work.

Membranes are key to understanding the compartmentalisation of the functions of a cell. The structure of a cell membrane is closely linked with its function and you will be looking at the link between the two.

By looking at the many different types of organelles within a cell you will discover both their structure, their function and how they work together. Key organelles in all cells are the mitochondria – so important they have their own DNA that enables them to divide independently. This is where the reactions of cellular respiration take place.

Plant cells have most if not all of the features found in animal cells – and a few of their own. You will be considering the structure and function of these features, which highlight at a cellular level some of the major differences between plants and animals.

All the maths you need

- Carry out calculations using numbers in standard and ordinary form (*e.g. use of magnification*)
- Use expressions in standard form (*e.g. when applied to areas such as the size of organelles*)
- Use scales for measuring (*e.g. graticule to measure size of cells*)
- Make order of magnitude calculations (*e.g. use and manipulate the magnification formula: magnification = size of image/size of real object*)
- Use and manipulate equations, including changing the subject of an equation (*e.g. magnification*)
- Calculate the circumference, surface areas and volumes of regular shapes (*e.g. calculate the surface area or volume of a cell*)

What will I study later?

- The details of the ultrastructure of prokaryotic cells including the organelles such as the nucleoid, plasmids, 70S ribosomes and bacterial cell walls
- The difference between Gram-positive and Gram-negative bacterial cell walls – and why it matters
- The structure of viruses – how they differ from all other living organisms and how they reproduce in other cells
- The eukaryotic cell cycle with the three stages of interphase, mitosis and cytokinesis
- Mitotic cell division producing two identical daughter cells and the importance of mitotic division
- Meiotic cell division producing four haploid gametes
- Meiotic cell division as a source of genetic variation through recombination
- Gametogenesis and fertilisation in mammals and plants
- Specialised blood cells
- Specialised cells making up xylem and phloem tissue in plants
- The importance of the structural features of mitochondria in cellular respiration and chloroplasts n photosynthesis (A level)
- The cells of the immune system (A level)
- Epigenetics and cell differentiation (A level)

What have I studied before?

- How the main sub-cellular structures of eukaryotic cells (plants and animals) are related to their functions, including the nucleus, mitochondria, chloroplasts and cell membranes
- How electron microscopy has increased our understanding of sub-cellular structures
- The main differences between animal and plant cells
- Certain specialised cells, e.g. neurones, red blood cells

What will I study in this chapter?

- The way specimens are magnified by the light microscope and the electron microscope
- The way specimens are prepared for the light microscope (including staining) and the electron microscope
- The difference between magnification and resolution
- The details of the ultrastructure of eukaryotic cells related to the functions of the membrane-bound organelles
- The importance and structure of cell membranes
- The main membrane-bound organelles of animal eukaryotic cells, including the nucleus, nucleolus, 80S ribosomes, rough and smooth endoplasmic reticulum, mitochondria, centrioles, lysosomes and Golgi apparatus
- The main differences in structure between animal and plant cells, including cell walls, chloroplasts, vacuoles and tonoplasts

Cells are discussed in the media on an almost daily basis in relation to topics such as cancer, stem cells and DNA testing. However, in spite of the fact that we have known about cells for over 300 years, most people have only a vague idea about what they are and how they function.

Discovering cells

Robert Hooke (1635–1703), an English architect and natural philosopher, designed and put together one of the first working optical microscopes. His observations were published in his book *Micrographia* in 1665. Amongst the many objects he examined were thin sections of cork, made up of tiny, regular compartments that he called cells, as they reminded him of the monks' cells in a monastery. In 1676 Anton van Leeuwenhoek (1632–1723), a Dutch draper who ground lenses in his spare time to check the weave of his fabrics, used his lenses to observe a wide variety of living unicellular organisms in drops of water, which he called 'animalcules'. At the same time the English plant scientist Nehemiah Grew (1641–1712) was one of the first scientists to publish accurate drawings of 'tissues'. By the 1840s we understood that cells are the basic units of life, an idea that was first expressed by Matthias Schleiden (1804–1881) and Theodor Schwann (1810–1882) in their cell theory of 1839. Cell theory is now accepted as a unifying concept in biology. It states that cells are a fundamental unit of structure, function and organisation in all living organisms. Improvements in the quality of lenses, new staining techniques and the introduction of new technologies such as electron and confocal microscopes, have allowed us to see cells in increasing detail and so develop our understanding of both their structure and function.

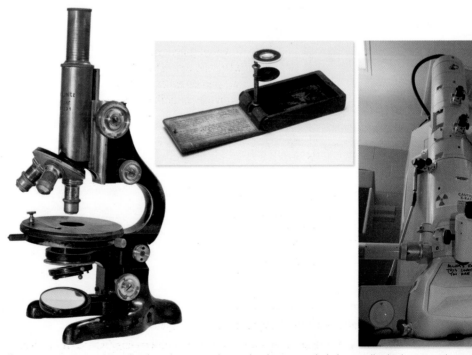

fig A As microscopes have developed, more and more has been revealed about cells, the key to understanding biology.

Microscopes

We can see some cells easily with the naked eye, for example the ovum in an unfertilised bird's egg is a single cell. But we need some kind of magnification to enable us to see most cells.

The **light microscope** or **optical microscope** has been the main tool for observing cells over the years and it is still widely used. A good light microscope can magnify to 1500 times and still give a clear image. At this magnification an average person would appear to be 2.5 km tall.

Since the mid-twentieth century the **electron microscope** has given scientists an even greater insight into the inner workings of cells. An electron microscope can give a magnification of up to 500 000 times, making an average person appear over 830 km tall!

Did you know?

Magnification and **resolution** are the two features of any microscope that determine how clear the image is:

- Magnification is a measure of how much bigger the image you see is than the real object, e.g. ×40, ×1000 or ×500 000.
- Resolution or **resolving power** is a measure of how close together two objects can be before we see them as one. For example, the resolution of the naked eye is around 0.1 mm. Two objects closer together than 0.1 mm cannot be seen as separate objects. The resolution of a light microscope is around 0.2 μm (200 nm), and the resolution of an electron microscope is around 0.1–1 nm.

The light microscope

A specimen or thin slice of biological material is placed on the stage of a light microscope (see **fig B**) and illuminated from underneath, either by sunlight reflected with a mirror or by a built-in light source. The objective lens produces a magnified and inverted image, which the eyepiece lens focuses at the eye. The total magnification of the specimen is calculated:

magnification of objective lens	×	magnification of eyepiece lens	=	total magnification
e.g. ×10	×	×10	=	×100

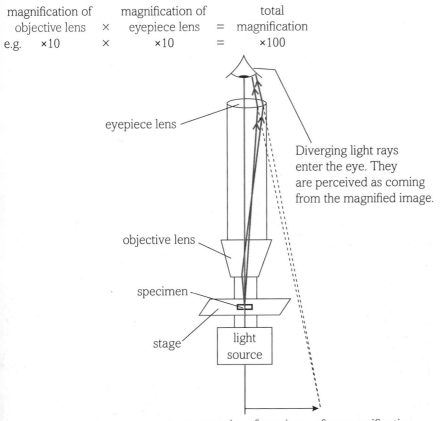

eyepiece lens

Diverging light rays enter the eye. They are perceived as coming from the magnified image.

objective lens

specimen

stage

light source

apparent size of specimen after magnification

fig B Light passes through the specimen and on through the lenses to give an image that is magnified and upside down.

WORKED EXAMPLE

Using the equation:

image size = actual size × magnification

You can work out the size of a specimen by measuring it under the microscope, as long as you always record the magnification you are using.

For example, the diameter of a cell measured under the light microscope at magnification ×400 is 1 mm.

Multiply 1 by 1000 to convert mm to µm:

$$1 \times 1000 = 1000$$

$$\frac{\text{image size}}{\text{magnification}} = \text{actual size}$$

$$\frac{1000}{400} = 2.5 \text{ µm}$$

You can look at living organisms, tissues and cells under the light microscope. However, most of the specimens will be dead, stained, specially preserved and sectioned (very thinly sliced) before they are mounted on a slide. The staining is used to make it easier to identify particular types of cell, or particular parts of the cells, under the microscope. Some of the stains you may come across include:

- haematoxylin – stains the nuclei of plant and animal cells purple, blue or brown
- methylene blue – stains the nuclei of animal cells blue
- acetocarmine – stains the chromosomes in dividing nuclei in both plant and animal cells
- iodine – stains starch-containing material in plant cells blue-black.

There are big advantages to using light microscopes, but there are some disadvantages too:

Advantages of the light microscope	Disadvantages of the light microscope
• Can see living plants and animals, or parts of them, directly. This is useful in itself and allows you to compare prepared slides with living tissue. • Relatively cheap so are available in schools and universities, hospitals, industrial labs and research labs. • Relatively light and portable so we can use them almost anywhere, e.g. identifying malaria in the field.	• Preservation and staining tissue can produce artefacts in the tissues being observed, so what we see may be the result of preparation rather than a true representation of the living tissue. • Limited powers of resolution and magnification.

Developments including the confocal microscope mean the information we can get from light microscopes continues to increase.

The electron microscope

The electron microscope uses a beam of electrons to form an image. The electrons are scattered by the specimen in much the same way as light is scattered in the light microscope. In an electron microscope, the electrons effectively behave like light waves with a very tiny wavelength. Electromagnetic or electrostatic lenses focus the electron beam to form an image. Resolving power increases as the wavelength gets smaller, so the electron microscope can resolve detail down to less than 0.00001 µm, about 10 000 times better than the light microscope.

For the electron microscope to work, the specimens have to be in a vacuum, so they are always dead. The preparation of a specimen for the electron microscope is a very complex process that may involve chemical preservation, freeze drying, freeze fracturing, removing the water (dehydration), embedding, sectioning and mounting on a metal grid. Specimens for electron microscopy are often stained using heavy metal ions such as lead and uranium. This is not to identify particular tissues, but to improve the scattering of the electrons and make greater contrast in the image, making it clearer and easier to interpret. The image is displayed on a monitor or computer screen.

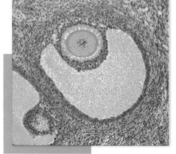

fig C A good light micrograph of tissue with staining that shows up the different types of cells can provide us with a lot of information, as demonstrated by this section through ovarian tissue.

Be very clear about the difference between magnification and resolution.

Make sure you are able to calculate the size, magnification or image size of any specimen.

There are two main types of electron micrographs. **Transmission electron micrographs (TEMs)** are two-dimensional (2D) images like those from a light microscope. **Scanning electron micrographs (SEMs)** have a lower magnification, but are three-dimensional (3D) and can be very striking. Sometimes electron micrographs are given false colours to make it easier to identify the different cells, but these are not stains. They are added after the image has been taken.

There are big advantages to using electron microscopes, but there are some disadvantages too:

Advantages of the electron microscope	Disadvantages of the electron microscope
• Huge powers of magnification and resolution. Many details of cell structure have been seen for the first time since they were developed.	• All specimens are examined in a vacuum – air would scatter the electrons and make the image of the tissue fuzzy – so it is impossible to look at living material.
	• Specimens undergo severe treatment that is likely to result in artefacts. Preparing specimens for the electron microscope is very skilled work.
	• Extremely expensive.
	• Large, have to be kept at a constant temperature and pressure and need to maintain an internal vacuum. Relatively few scientists outside research laboratories have easy access to such equipment.

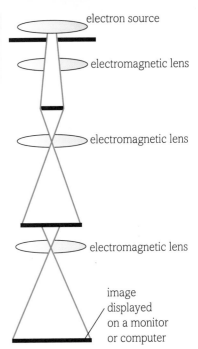

fig D A beam of electrons passes through the specimen and on through electromagnetic or electrostatic lenses to give a greatly magnified image.

(labels on fig D: electron source; electromagnetic lens; electromagnetic lens; electromagnetic lens; image displayed on a monitor or computer)

Questions

1 Why is high magnification alone not enough to give us biological details of cells?

2 Both light and electron micrographs can be brightly coloured. Explain the differences and similarities between the way colour is used in light and electron microscopy.

3 A student measured the diameter of three cells of the same type under the microscope. Measurement 1 was taken with a magnification of ×40, and measurements 2 and 3 with a magnification of ×100. Work out the mean diameter of the cells.

Measurement 1 = 5 mm   Measurement 2 = 12 mm   Measurement 3 = 1 mm

Key definitions

A **light microscope (optical microscope)** is a tool that uses a beam of light and optical lenses to magnify specimens up to 1500 times life size.

An **electron microscope** is a tool that uses a beam of electrons and magnetic lenses to magnify specimens up to 500 000 times life size.

Magnification is a measure of how much bigger the image you see is than the real object.

Resolution (resolving power) is a measure of how close together two objects can be before they are seen as one.

Transmission electron micrographs (TEMs) are micrographs produced by the electron microscope that give 2D images like those from a light microscope, but magnified up to 500 000 times.

Scanning electron micrographs (SEMs) are micrographs produced by the electron microscope that have a lower magnification than TEMs, but produce a 3D image.

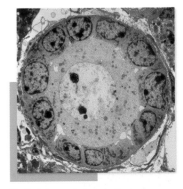

fig E A transmission electron micrograph of a cell gives you much more detailed information than the light micrograph in **fig C**.

fig F Scanning electron micrographs open up a whole new, 3D world of biology on a small scale.

By the end of this section, you should be able to...

● define eukaryotic and prokaryotic cells

● recognise that cell membranes are common to eukaryotes and prokaryotes

● explain how the structure and properties of phospholipids relate to their function in cell membranes

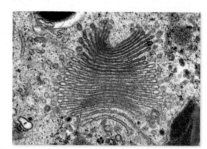

fig A One of the functions of this stack of membranes in a plant cell (the Golgi apparatus) is to package secretions into vesicles. It is just one of the membrane-rich organelles you can find in a eukaryotic cell.

Most of the familiar organisms in the world around you have the same sort of cells. Animals, plants, protoctists, algae and fungi have cells with the genetic material contained in a membrane-bound nucleus. The cells also contain a number of other membrane-bound **organelles** such as mitochondria and chloroplasts. These organisms are called the **eukaryotes**. But there is another ancient group of organisms, including the bacteria and cyanobacteria, known as the **prokaryotes**. They have cells of a very different type that lack much of the structure and organisation of the eukaryotic cells, but they do have a **cell surface membrane**. You will look at both eukaryotic and prokaryotic cells in this topic, but you will begin by studying the structure of the membranes that are common to all.

Membranes in cells

There are many membranes within cells, such as those that surround organelles like the nucleus and mitochondria. But the most obvious membrane is the cell surface membrane, also known as the outer cell membrane, which forms the boundary of all cells – controlling what passes into and out of the cell and allowing the fluids either side of it to have different compositions. Membranes within cells make it possible to have the right conditions for a particular reaction in one part of a cell and different conditions to suit other reactions elsewhere in the same cell.

Membranes perform many other functions too. Many chemical processes take place on membrane surfaces. For example, the reactions of respiration in eukaryotic cells take place on the inner mitochondrial membrane. Enzymes and any other factors are held closely together so that the reaction processes can proceed smoothly. The cell surface membrane must also be flexible to allow the cell to change shape very slightly as its water content changes, or quite dramatically, for example when a white blood cell engulfs a bacterium. Chemical secretions made by the cell are packaged into membrane bags known as **vesicles**, so some membranes must be capable of breaking and fusing together readily.

The structure of membranes

Our current model of the structure of membranes has been worked out over many years. The model developed as microscopy improved, from light to electron and then scanning electron microscopes. In time there may well be further refinements to the model presented here, but the overall picture seems unlikely to change dramatically. The membrane is made up mainly of two types of molecules – lipids and proteins – arranged in a very specific way.

The phospholipid bilayer

The lipids in the membrane are of a particular type called **polar lipids**. These are lipid molecules with one end joined to a polar group. Many of the polar lipids in the membrane are phospholipids, with a phosphate group forming the polar part of the molecule (see **Section 1.2.3**). With water or aqueous solutions on each side, phospholipid molecules form a bilayer with their hydrophilic heads pointing into the water while the hydrophobic tails stay protected in the middle. This structure is known as a unit membrane.

However, a simple lipid bilayer alone would not explain either the microscopic appearance of membranes or the way in which they behave. A simple lipid bilayer allows fat-soluble organic molecules to pass through it, but many vital chemicals needed in cells are ionic. Whilst these dissolve in water they cannot dissolve in or pass through lipids, even polar lipids. They can enter cells because the membrane consists not only of lipids, but also of proteins and other molecules.

The membrane proteins

The best model of a membrane we have today sees the basic bilayer of phospholipid as a fluid system, with many proteins and other molecules floating within it like icebergs whilst others are fixed in place (see **fig B**). The proportion of phospholipids containing unsaturated fatty acids (see **Section 1.2.3**) in the bilayer seems to affect how freely the moving proteins float about in the membrane. The more unsaturated fatty acids, the more fluid the membrane. Many of the proteins have a hydrophobic part, which is buried in the lipid bilayer, and a hydrophilic part, which can be involved in a variety of activities. Some proteins penetrate all the way through the lipid, while others only go part of the way through the bilayer.

One of the main functions of the membrane proteins is to help substances move across the membrane. The proteins can form pores or channels – some permanent, some temporary – that allow specific molecules to move through. Some of these channels can be open or shut, depending on conditions in the cell. These are known as **gated channels**. Some of the protein pores are active carrier systems using energy to move molecules, as you will see later. Others are simply gaps in the lipid bilayer that allow ionic substances to move through the membrane in both directions.

Proteins may act as specific receptor molecules – for example, making cells sensitive to a particular hormone. They may be enzymes, particularly on any internal cell membranes, to control reactions linked to that membrane. Some membrane proteins are glycoproteins, proteins with a carbohydrate part added to the molecule. These are very important on the surface of cells as part of the way cells recognise each other.

This model of the floating proteins in a lipid sea is known as the **fluid mosaic model** and was first proposed by S. Jonathan Singer (1924–) and Garth Nicholson (1943–) in 1972.

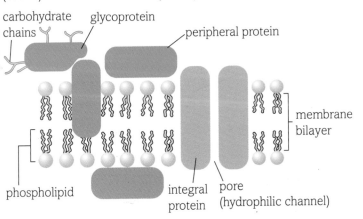

carbohydrate chains
glycoprotein
peripheral protein
membrane bilayer
phospholipid
integral protein
pore (hydrophilic channel)

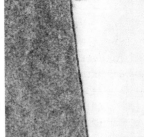

fig B Whether acting as the boundary of a cell or as part of its internal make-up, the complex structure of the membrane is closely linked to its wide variety of functions.

Did you know?

Evidence for the fluid mosaic model

Techniques such as X-ray diffraction and advanced electron microscopy have added to our knowledge of the structure of cell membranes, giving us more details of the layers, the pores and the carrier molecules.

However, even under the electron microscope, cell membranes are very small. Microscopes have helped scientists develop our current membrane model, but other techniques are also important. We can identify proteins that appear to have a specific function of transporting particular ions into or out of the cell through the membrane. Cystic fibrosis is a genetic disease that affects transport across the membranes of the glands, the digestive system, the respiratory system and the reproductive system. If an individual inherits a faulty allele (variant) from each parent, the protein needed to transport chloride ions across the membranes (cystic fibrosis transmembrane regulatory channel protein) does not form properly. This affects the movement of water out of the cells and leads to the formation of sticky mucus, which can lead to serious chest infections, digestive problems and infertility. Identifying protein channels in the membrane that have a very specific function that can be measured helps to confirm our model of cell membrane structure and function.

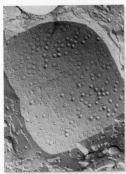

fig C The membrane pores through which mRNA leaves the nucleus are clearly visible in this freeze-etched electron micrograph of the nuclear membrane of a cell.

Questions

1 Summarise the main functions of membranes in cells.

2 Which kinds of molecule make up the structure of a membrane and how do their properties affect the properties of the membrane itself?

3 Discuss why the flexible structure of the cell membrane is well adapted for its functions in the cell.

Key definitions

Organelles are sub-cellular bodies found in the cytoplasm of cells.

Eukaryotes are a group of organisms with cells that have the genetic material contained in a membrane-bound nucleus and also contain a number of membrane-bound organelles such as mitochondria and chloroplasts.

Prokaryotes are a group of organisms including bacteria and blue-green algae (cyanobacteria) that have few organelles and do not have the genetic material contained in a membrane-bound nucleus.

The **cell surface membrane** is the membrane that forms the outer boundary of the cytoplasm of a cell and controls the movement of substances into and out of the cell.

Vesicles are membrane 'bags' that hold secretions made in cells.

Polar lipids are lipids with one end attached to a polar group, e.g. a phosphate group that makes one end of the molecule hydrophilic and one end hydrophobic.

Gated channels are protein channels through the lipid bilayer of a membrane that are opened or closed, depending on conditions in the cell.

The **fluid mosaic model** is the current model of the structure of the cell membrane including floating proteins forming pores, channels and carrier systems in a lipid bilayer.

Eukaryotic cells 1 – common cellular structures

By the end of this section, you should be able to...

● describe eukaryotic cells

● describe the ultrastructure of eukaryotic cells and the functions of organelles including the nucleus, nucleolus, mitochondria, centrioles and vacuoles

Most microscope images, apart from those of living material or from a scanning electron microscope, make cells appear flat and two-dimensional (2D). But cells are actually spheres, cylinders or asymmetrical three-dimensional (3D) shapes – so try to use your imagination when you look at cells and visualise them in three dimensions.

The characteristics of eukaryotic cells

In eukaryotic organisms such as animals, plants and fungi there is a very wide range of different types of cell, each with a different function. But there are certain cell features that turn up again and again, and we can put these together as a 'typical' plant or animal cell. Remember that this typical cell does not really exist, but acts as a useful guide to what to look for in any eukaryotic cell.

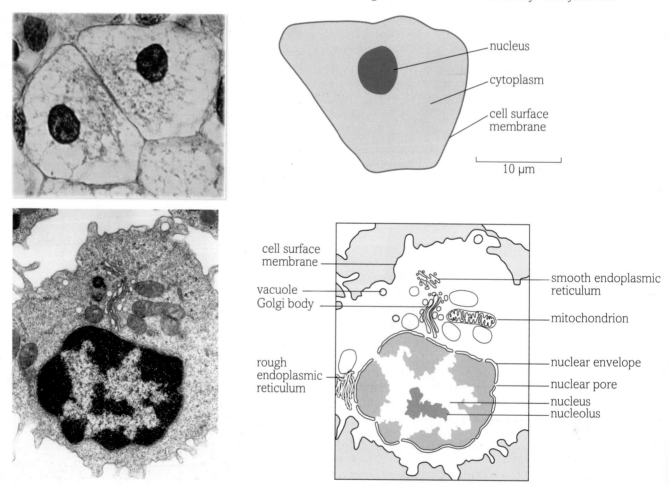

fig A These images show clearly how the introduction of the electron microscope increased our detailed knowledge and understanding of structures within cells.

The typical animal cell

A typical animal cell contains many things that are common to all eukaryotic cells, including plants and fungi. Inside the cell surface membrane is a jelly-like liquid called the **cytoplasm**, containing a **nucleus** – the two together are known as the **protoplasm**. The cytoplasm contains most of what is needed to carry out the day-to-day tasks of living, whilst the nucleus is vital to the long-term survival of the cell, because it contains the information needed to produce all the chemicals that make up the cell. This basic pattern gives rise to an enormous number of variations suited for the different functions that arise within the animal kingdom. The various parts of the cell have complex and detailed structures, which we can see more clearly when an electron microscope is used. The structures that can only be observed in detail using the electron microscope are known as the **ultrastructure** of the cell. The structure of each part of the cell relates closely to the job it has to do.

Membranes

Membranes in a cell are important both as an outer boundary to the cell and in the multitude of internal (**intracellular**) membranes. In **Section 2.1.2** you looked at the importance of cell membranes for controlling the movement of substances, but membranes inside the cell also have other functions. They localise enzymes in reaction pathways, for example respiration in mitochondria and photosynthesis in chloroplasts, and they compartmentalise chemicals, for example hydrolytic enzymes in lysosomes. You will learn more about membrane functions throughout this section as you consider the various structures that make up the cell.

The protoplasm

When the light microscope was the only tool biologists had to observe cells, they thought that the cytoplasm was a relatively structureless, clear jelly. But the electron microscope revealed the cytoplasm to be full of all manner of structures, known as organelles, some of which are described below.

The nucleus

The nucleus is usually the largest organelle in the cell (1–20 µm) and it can be seen with the light microscope. Electron micrographs show that the nucleus, which is usually spherical in shape, is surrounded by a double nuclear membrane containing holes or pores, known as the nuclear envelope. Chemicals can pass in and out of the nucleus through these pores so that the nucleus can control events in the cytoplasm. Inside the nuclear envelope are two main substances, nucleic acids and proteins. The nucleic acids are deoxyribonucleic acid (DNA) and ribonucleic acid (RNA) (see **Chapter 1.3**).

When the cell is not actively dividing, the DNA is bonded to the protein to form **chromatin**, which looks like tiny granules. Also in the nucleus there is at least one **nucleolus** – an extra-dense area of almost pure DNA and protein. The nucleolus is involved in the production of ribosomes. Recent research also suggests that the nucleolus plays a part in the control of cell growth and division.

Mitochondria

The name **mitochondrion** simply means 'thread granule' and describes the tiny rod-like structures that are 1 µm wide by up to 10 µm long, seen in the cytoplasm of almost all eukaryotic cells under the light microscope. In recent years, by using the electron microscope we have been able to understand not only their complex structure, but also their vital functions.

The mitochondria are the 'powerhouses' of the cell. Here, in a series of complicated biochemical reactions, simple molecules are oxidised in the process of cellular respiration, producing ATP (see **Chapter 1.3**) that can be used to drive the other functions of the cell and indeed the organism. The number of mitochondria present can give you useful information about the functions of a cell. Cells that require very little energy, for example white fat storage cells, have very few mitochondria. Any cell with an energy demanding function, for example muscle cells or cells that carry out a lot of active transport such as liver cells, will contain large numbers of mitochondria.

An outer and inner membrane surround the mitochondria. They also contain their own genetic material, so that when a cell divides, the mitochondria replicate themselves under the control of the nucleus. This mitochondrial DNA is part of the whole genome of the organism.

Mitochondria have an internal arrangement adapted for their function (see **fig B**). The inner membrane is folded to form **cristae**, which give a very large surface area, surrounded by a fluid matrix. This structure is closely integrated with the events in cellular respiration that take place in the mitochondrion (see **Book 2 Sections 5.1.3** and **5.1.4**). Backed by evidence that shows that mitochondria have their own DNA, scientists think that mitochondria and chloroplasts originated as symbiotic **eubacteria** living inside early cells. Over millions of years of evolution they have become an integral part of the cell (see **Section 3.1.5**). This is the **endosymbiotic theory** of the evolution of eukaryotic cells.

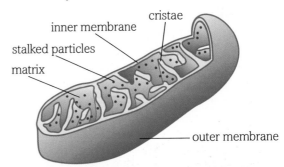

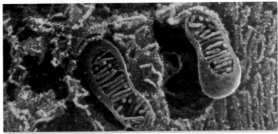

fig B The 3D structure of the mitochondria (blue) is closely related to their functions in cellular respiration.

The centrioles

In each cell there is usually a pair of **centrioles** near the nucleus (see **fig C**). Each centriole is made up of a bundle of nine tubules and is about 0.5 μm long by 0.2 μm wide. The centrioles are involved in cell division. When a cell divides, the centrioles pull apart to produce a **spindle** of microtubules that are involved in the movement of the chromosomes, as you will see later in this chapter.

(a)

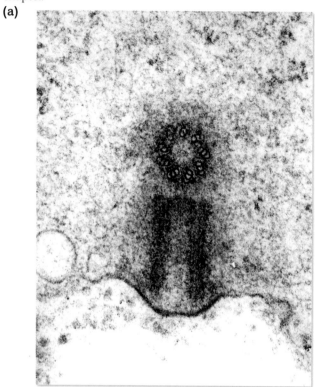

(b)

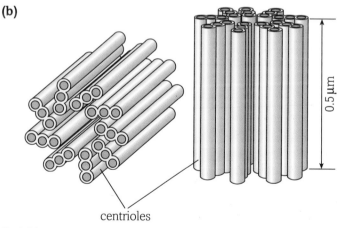

centrioles

fig C (a) Transition electron micrograph of centrioles; and (b) Diagram of centrioles.

The cytoskeleton

A cellular skeleton may seem a contradiction in terms, yet work in recent years has shown that a **cytoskeleton** is a feature of all eukaryotic cells. It is a dynamic, 3D web-like structure that fills the cytoplasm (see **fig D**). It is made up of **microfilaments**, which are protein fibres, and **microtubules**, tiny protein tubes about 20 nm in diameter. Microtubules are found both singly and in bundles throughout the cytoplasm. These microtubules consist mainly of the globular protein tubulin. The cytoskeleton performs several functions. It gives the cytoplasm structure and keeps the organelles in place. Many of the proteins in the microfilaments are related to actin and myosin, the contractile proteins in muscle, and the cytoskeleton is closely linked with cell movements and transport within cells. Recent research has shown that a cytoskeleton is also a feature of some prokaryotic cells.

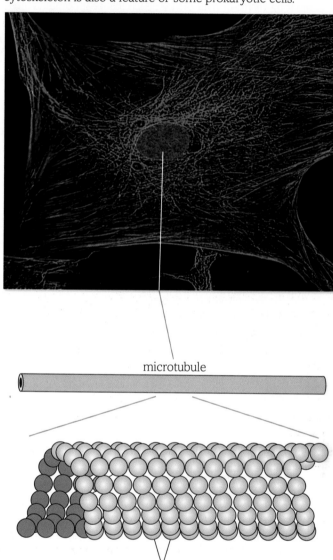

microtubule

tubulin sub-units

fig D The cytoskeleton forms a tangled web of structural and contractile fibres that hold the organelles in place and enable cell movement to occur.

Vacuoles

Vacuoles are not a permanent feature in animal cells. These membrane-lined enclosures are formed and lost as needed. Many simple animals make food vacuoles around the prey they engulf. White blood cells in higher animals form similar vacuoles around engulfed pathogens. **Contractile vacuoles** are an important feature in simple animals that live in fresh water because they allow the water content of the cytoplasm to be controlled. But in spite of these examples, vacuoles are not a major feature of animal cells and permanent vacuoles are never seen.

Learning tips

Remember that different types of electron microscopy provide very different types of information:

- The scanning EM can show intact organelles, allowing detailed measurements of the outer dimensions to be taken or it can take 3D images along fracture lines.
- The transmission EM provides clear images of the internal structures of the organelles.

Together the information is useful to produce a detailed image of the ultrastructure of a cell.

Questions

1 What is the role of the cytoskeleton in the cytoplasm and why has its importance only recently been recognised?

2 Explain the importance of organelles in eukaryotic cells.

3 Look at the different images that result from transmission and scanning electron microscopes in **Chapter 2** and describe how they differ. Suggest the advantages of each type of image and give examples where each would be more appropriate to use.

Key definitions

Cytoplasm is a jelly-like liquid that makes up the bulk of the cell and contains the organelles.

The **nucleus** is an organelle containing the nucleic acids DNA (the genetic material) and RNA, as well as protein, surrounded by a nuclear envelope with pores.

Protoplasm is the cytoplasm and nucleus combined.

The **ultrastructure** is the detailed organisation of the cell, only visible using the electron microscope.

Intracellular means inside the cell.

Chromatin is the granular combination of DNA bonded to protein found in the nucleus when the cell is not actively dividing.

A **nucleolus** is an extra dense area of almost pure DNA and protein found in the nucleus involved in the production of ribosomes and control of growth and division.

Mitochondria are rod-like structures with inner and outer membranes that are the site of aerobic respiration.

Cristae are the infoldings of the inner membrane of the mitochondria which provide a large surface area for the reactions of aerobic respiration.

Eubacteria are true bacteria (prokaryotic organisms).

The **endosymbiotic theory** is a theory that suggests that mitochondria and chloroplasts originated as independent prokaryotic organisms that began living symbiotically inside other cells as endosymbionts.

Centrioles are bundles of tubules found near the nucleus and involved in cell division by the production of a spindle of microtubules that move the chromosomes to the ends of the cell.

A **spindle** is a set of overlapping protein microtubules running the length of the cell, formed as the centrioles pull apart in mitosis and meiosis.

The **cytoskeleton** is a dynamic, 3D web-like structure made up of microfilaments and microtubules that fills the cytoplasm and gives it structure, keeping the organelles in place and enabling cell movements and transport within the cell.

Microfilaments are protein fibres that make up part of the structure of the cytoskeleton.

Microtubules are tiny protein tubes about 20 nm in diameter that make up part of the structure of the cytoskeleton.

A **vacuole** is a fluid-filled cavity within the cytoplasm of a cell surrounded by a membrane.

Contractile vacuoles are vacuoles that can fill and empty to help control the concentration of the cytoplasm of simple freshwater animals.

By the end of this section, you should be able to...

● describe the ultrastructure of eukaryotic cells and the functions of organelles including the rough and smooth endoplasmic reticulum, 80S ribosomes, Golgi apparatus and lysosomes

The cytoplasm of the cell contains the **endoplasmic reticulum (ER)**, a three-dimensional (3D) network of cavities bounded by membranes. The electron microscope reveals that some of the cavities are sac-like and some are tubular, and that the ER spreads extensively through the cytoplasm. The ER network links with the membrane around the nucleus, and makes up a large part of the transport system within a cell as well as being the site of synthesis of many important chemicals. It has been calculated that $1\,cm^3$ of liver tissue contains about $11\,m^2$ of endoplasmic reticulum. Electron microscopes also helped scientists to work out the functions of the endoplasmic reticulum, by showing up the different forms – the rough and the smooth endoplasmic reticulum.

Another useful technique is to provide cells with radioactively labelled chemicals that are building blocks for specific modules, for example labelled amino acids for the synthesis of proteins, and then find out where they appear in the cell. The labelled products can be tracked using microscopy. Another method of locating them is to break the cells open and then spin the contents in a centrifuge. The different parts of the cell can be separated out and the regions containing the radioactively labelled substances identified.

80S and 70S ribosomes

In **Section 1.3.5** you met ribosomes, the organelles on which protein synthesis takes place in the cytoplasm of the cell. Ribosomes are made from ribosomal RNA and protein, and consist of a large subunit and a small subunit. The main type of ribosomes in eukaryotic cells are **80S ribosomes**. The 'S' stands for Svedberg, a unit used to measure how quickly particles settle in a centrifuge. The rate of sedimentation depends on the size and shape of the particle. When 80S ribosomes are broken into their two units, they are made up of a 40S small subunit and a 60S large subunit. The ratio of RNA : protein in 80S ribosomes is 1 : 1.

However, eukaryotic cells also contain another type of ribosome. Scientists have discovered **70S ribosomes** in the mitochondria, and in the chloroplasts of plant cells. These ribosomes are usually found in prokaryotic cells (bacteria and cyanobacteria). They are made up of a small 30S subunit and a larger 50S subunit and the ratio of RNA : protein in 70S ribosomes is 2 : 1.

These 70S ribosomes are reproduced in the mitochondria and chloroplasts independently when a cell divides. This is seen as good evidence for the endosymbiotic theory that mitochondria and chloroplasts evolved from bacteria caught inside eukaryotic cells very early on in the process of evolution.

Rough and smooth endoplasmic reticulum

Electron micrographs show that much of the outside of the endoplasmic reticulum membrane is covered with granules, which are 80S ribosomes, so this is known as **rough endoplasmic reticulum (RER)** (see **fig A**). The function of the ribosomes is to make proteins and the RER isolates and transports these proteins once they have been made. Some proteins, such as digestive enzymes and hormones, are not used inside the cell that makes them, so they have to be secreted without interfering with the cell's activities. This is an example of **exocytosis**.

Many other proteins are needed within the cell. The RER has a large surface area for the synthesis of all these proteins, and it stores and transports them both within the cell and from the inside to the outside. Cells that secrete materials, such as those producing the digestive enzymes in the lining of the gut, have a large amount of RER.

Not all endoplasmic reticulum is covered in ribosomes (see **fig A**). **Smooth endoplasmic reticulum (SER)** is also involved in synthesis and transport, but in this case of steroids and lipids. For example, lots of SER is found in the testes, which make the steroid hormone testosterone, and in the liver, which metabolises cholesterol amongst other lipids. The amount and type of endoplasmic reticulum in a cell give an idea of the type of job the cell does.

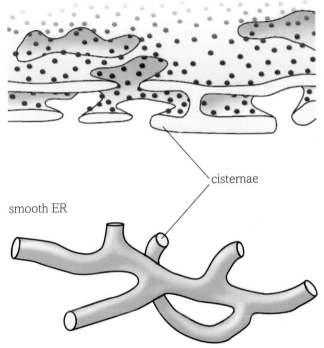

fig A Rough and smooth endoplasmic reticulum. Smooth ER is more tubular than rough ER and also lacks ribosomes on the surface.

The Golgi apparatus

Under the light microscope the **Golgi apparatus** looks like a rather dense area of cytoplasm. An electron microscope reveals that it is made up of stacks of parallel, flattened membrane pockets called cisternae, formed by vesicles from the endoplasmic reticulum fusing together (see **Section 2.1.2, fig A**).

The Golgi apparatus has a close link with, but is not joined to, the RER. It has taken scientists a long time to discover exactly what the Golgi apparatus does. Materials have been radioactively labelled and tracked through the cell to try and find out exactly what goes on inside it. Proteins are brought to the Golgi apparatus in vesicles that have pinched off from the RER where they were made. The vesicles fuse with the membrane sacs of the Golgi apparatus and the protein enters the Golgi stacks. As the proteins travel through the Golgi apparatus they are modified in various ways.

Carbohydrate is added to some proteins to form glycoproteins such as mucus. The Golgi apparatus also seems to be involved in producing materials for plant and fungal cell walls and insect cuticles. Some proteins in the Golgi apparatus are digestive enzymes. These may be enclosed in vesicles to form an organelle known as a **lysosome**. Alternatively, enzymes may be transported

through the Golgi apparatus and then in vesicles to the cell surface membrane where the vesicles fuse with the membrane to release extracellular digestive enzymes. The Golgi apparatus was first reported over 100 years ago, in April 1898. The flattened stack of membranes was observed by the Italian scientist Camillo Golgi (1843–1926) through a light microscope. For more than 50 years scientists argued over its function. Some thought it was an artefact from the process of fixing and staining during tissue preparation. The arrival of the electron microscope in the 1950s allowed the detailed structure of the Golgi apparatus to be seen.

The electron microscope has been central in showing details of the internal structure of the Golgi apparatus. In addition, a number of techniques have been developed that have allowed more detailed understanding. The most important of these has been the process of labelling specific enzymes so they can be seen using the electron microscope. The inner areas of the Golgi apparatus, nearer to the RER, have been shown to be very rich in enzymes that modify proteins in various ways. This is where most enzymes or membrane proteins are converted into the finished product. In contrast, in the outer regions of the Golgi apparatus you find lots of finished protein products, but not many of the enzymes that make them. The movement of cell membrane proteins through the Golgi apparatus is very complex. Areas of the protein that need to be on the outside of the cell membrane, such as receptor binding sites, are orientated by the Golgi apparatus so that when they arrive at the membrane they are inserted facing in the right direction.

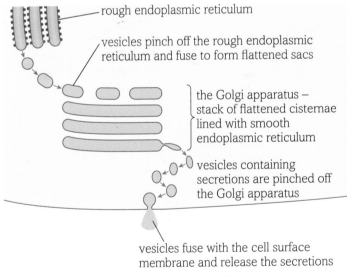

rough endoplasmic reticulum

vesicles pinch off the rough endoplasmic reticulum and fuse to form flattened sacs

the Golgi apparatus – stack of flattened cisternae lined with smooth endoplasmic reticulum

vesicles containing secretions are pinched off the Golgi apparatus

vesicles fuse with the cell surface membrane and release the secretions

fig B The Golgi apparatus takes proteins from the RER, assembles and packages them and then transports them to where they are needed. This may be the surface of the cell or different regions inside it.

Lysosomes

Food taken into the cell of single-celled protoctists such as *Amoeba* must be broken down into simple chemicals that can then be used. Organelles in the cells of your body that are worn out need to be destroyed. These jobs are the function of the lysosomes. The word lysis, from which they get their name, means 'breaking down'. Lysosomes appear as dark, spherical bodies in the cytoplasm of most cells and they contain a powerful mix of digestive enzymes. They frequently fuse with each other and with a membrane-bound vacuole containing either food or an obsolete organelle. Their enzymes then break down the contents into molecules that can be reused. A lysosome may fuse with the outer cell membrane to release its enzymes outside the cell as extracellular enzymes, for example to destroy bacteria or in digestion.

Lysosomes can also self-destruct. If an entire cell is wearing out, needs to be removed during development, has a mutation or is under stress, its lysosomes may rupture, releasing their enzymes to destroy the entire contents of the cell. This programmed, controlled cell death is known as **apoptosis**.

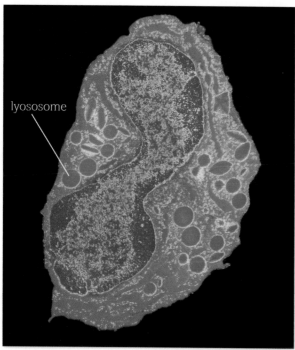

lyososome

fig C Good microscopic evidence of lysosomes helped scientists work out their functions in the cell.

Did you know?

Apoptosis and disease

Apoptosis or programmed cell death is vital to the maintenance of a healthy body. Lysosomes rupture and their enzymes are released to kill cells that are old and coming to the end of their healthy life, or cells that need to be removed during development, for example the webbing that initially forms between the fingers and toes of a fetus in the uterus. Lysosomes may also destroy cells in which the DNA replication system is not functioning properly. But if apoptosis stops working properly – if too many cells are destroyed, or not enough lysosomes rupture so that cell death no longer takes place – this can have serious consequences for your health. For example, cancer is often thought of as a disease of uncontrolled cell growth. But scientists are increasingly convinced that uncontrolled growth is not the whole story. Cancer cells also fail to die by apoptosis. As a result they propagate the genetic mutations that allow them to reproduce uncontrollably. Excessive apoptosis also causes problems. It leads to the damage seen in the heart after a heart attack, and is linked to the death of T killer cells in HIV/AIDS. This is covered in more detail in **Book 2**. The excessive rupturing of lysosomes may also be involved in autoimmune diseases such as rheumatoid arthritis, when cartilage tissue in joints self-destructs, and possibly in other conditions such as osteoporosis and retinitis pigmentosa.

Questions

1 What type of questions would scientists have asked when they set out to investigate the functions of the endoplasmic reticulum, and how might they have set about finding the answers?

2 Describe the role of the RER and the Golgi body in the production of both intracellular and extracellular enzymes, and explain the importance of packaging products within a cell.

3 Why is it important that apoptosis does not occur more or less than it should? Investigate examples of diseases that are caused at least in part by apoptosis.

Key definitions

The **endoplasmic reticulum** is a 3D network of membrane-bound cavities in the cytoplasm that links to the nuclear membrane and makes up a large part of the cellular transport system as well as playing an important role in the synthesis of many different chemicals.

80S ribosomes are the main type of ribosome found in eukaryotic cells, consisting of ribosomal RNA and protein, made up of a 60S and 40S subunit. They are the site of protein synthesis.

70S ribosomes are found in the mitochondria and chloroplasts of eukaryotic cells and in prokaryotic organisms.

Rough endoplasmic reticulum (RER) is endoplasmic reticulum that is covered in 80S ribosomes and which is involved in the production and transport of proteins.

Exocytosis is the energy-requiring process by which a vesicle fuses with the cell surface membrane so the contents are released to the outside of the cell.

Smooth endoplasmic reticulum (SER) is a smooth tubular structure similar to RER, but without the ribosomes, which is involved in the synthesis and transport of steroids and lipids in the cell.

The **Golgi apparatus** consists of stacks of membranes that modify proteins made elsewhere in the cell and package them into vesicles for transport, and also produce materials for plant cell walls and insect cuticles.

A **lysosome** is an organelle full of digestive enzymes used to break down worn out cells or organelles, or digest food in simple organisms.

Apoptosis is programmed cell death – the breakdown of worn out, damaged or diseased cells by the lysosomes.

By the end of this section, you should be able to...

● describe the ultrastructure of the cell wall in eukaryotic cells and relate its structure to its functions

Plants, like animals, are eukaryotes. A typical plant cell has many features in common with a typical animal cell (see **Sections 2.1.3** and **2.1.4**). They have many membranes and contain cytoplasm and a nucleus. Rough and smooth endoplasmic reticulum spread throughout the cytoplasm, along with an active Golgi apparatus. Mitochondria produce ATP, which is as vital to the working of the plant cell as it is to the animal cell. However, there are several quite fundamental differences between plant and animal cells. They contain several kinds of organelle that are not found in animal cells, including permanent vacuoles and chloroplasts.

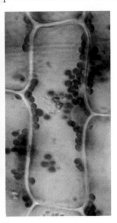
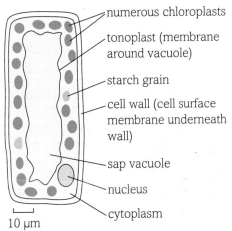

numerous chloroplasts

tonoplast (membrane around vacuole)

starch grain

cell wall (cell surface membrane underneath wall)

sap vacuole

nucleus

cytoplasm

10 μm

(a) a light micrograph and drawing of a plant cell ×250

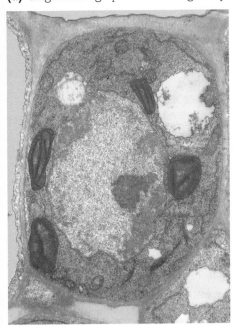
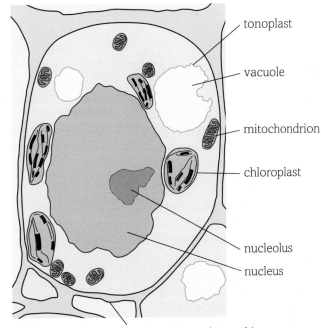

tonoplast

vacuole

mitochondrion

chloroplast

nucleolus

nucleus

cell wall (with cell surface membrane beneath)

(b) an electron micrograph and drawing of a plant cell ×5000

fig A The light microscope gives us the major features of a plant cell; the electron microscope reveals many more details.

The plant cell wall

Animal cells can be almost any shape. Plant cells tend to be more regular and uniform in their appearance. This is largely because each cell is bounded by a **cell wall**. You can visualise a plant cell as a jelly-filled balloon inside a shoe box. The cell wall (shoe box) is an important feature that gives plants their strength and support. It is made up largely of insoluble cellulose (see **Section 1.2.2**). The plant cell wall is usually freely permeable to everything that is dissolved in water – it does not act as a barrier to substances getting into the cell. However, the cell wall can become impregnated with **suberin** in cork tissues, or with **lignin** to produce wood. These compounds affect the permeability of the cell wall so that water and dissolved substances cannot pass through it.

fig B These cellulose microfibrils are made up of thousands of cellulose chains held together by hydrogen bonds. Their orientation and packing changes from primary to secondary cell walls, affecting both flexibility and strength.

The plant cell wall consists of several layers. The **middle lamella** is the first layer to form when a plant cell divides into two new cells. It is made largely of **pectin**, a polysaccharide that acts like glue and holds the cell walls of neighbouring plant cells together. Pectin has lots of negatively charged carboxyl (–COOH) groups and these combine with positive calcium ions to form calcium pectate. This binds to the cellulose that forms on either side. The cellulose microfibrils and the matrix build up on either side of the middle lamella. To begin with, these walls are very flexible, with the cellulose microfibrils all orientated in a similar direction. They are known as **primary cell walls**. As the plant ages, secondary thickening may take place. A **secondary cell wall** builds up, with the cellulose microfibrils laid densely at different angles to each other. This makes the composite material much more rigid. Hemicelluloses harden it further. In some plants, particularly woody perennials, lignin is then added to the cell walls to produce wood, which makes the structure even more rigid. Within the structure of a plant there are many long cells with cellulose cell walls that have been heavily lignified. These are known as **plant fibres** and people use them in many different ways including clothing, building material, ropes and paper.

Plasmodesmata

In spite of being encased in cellulose cell walls, plant cells seem to be in close communication with each other.

Intercellular exchanges seem to take place through special cytoplasmic bridges between the cells known as **plasmodesmata** (see **fig C**). The plasmodesmata appear to be produced as the cells divide – the two cells do not separate completely, and threads of cytoplasm remain between them. These threads pass through gaps in the newly formed cell walls and signalling substances can pass from one cell to another through the cytoplasm. The interconnected cytoplasm of the cells is known as the **symplast**. Scientists are still working hard to discover exactly how plant cells communicate through plasmodesmata. One clear piece of evidence showing that these intercellular junctions are vital in the life of plants comes from work with plant grafts. If we graft a rose onto a hardy root stock, the graft tissue only starts healthy cell division and growth once plasmodesmata bridges are established between the host tissue and the graft tissue.

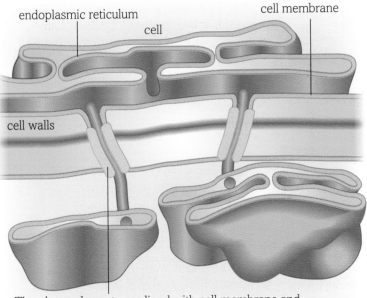

endoplasmic reticulum

cell

cell membrane

cell walls

The plasmodesmata are lined with cell membrane and molecules pass freely from cell to cell through these canals.

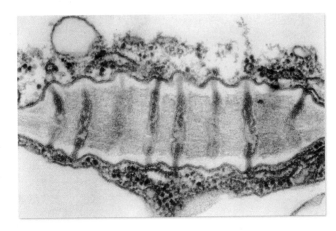

fig C Plasmodesmata provide a route for communication between plant cells, but scientists are still trying to find out exactly how it works.

Questions

1 What role do cell walls play in the structure of a plant, and how is their structure related to their function?

2 How does the plant cell wall change as the cell grows and develops, and how does this affect the cell?

3 Explain why plasmodesmata are an important feature of plant cell structure.

Key definitions

A **cell wall** is a freely permeable wall around plant cells, made mainly of cellulose.

Suberin is a chemical that impregnates cellulose cell walls in cork tissues and makes them impermeable.

Lignin is a chemical that impregnates cellulose cell walls in wood and makes it impermeable.

The **middle lamella** is the first layer of the plant cell wall to be formed when a plant cell divides, made mainly of calcium pectate (pectin) that binds the layers of cellulose together.

Pectin is a polysaccharide that holds cell walls of neighbouring plant cells together and is part of the structure of the primary cell wall.

The **primary cell wall** is the first very flexible plant cell wall to form, with all the cellulose microfibrils orientated in a similar direction.

The **secondary cell wall** is the older plant cell wall in which the cellulose microfibrils have built up at different angles to each other making the cell wall more rigid.

Plant fibres are long cells with cellulose cell walls that have been heavily lignified so they are rigid and very strong.

Plasmodesmata are cytoplasmic bridges between plant cells that allow communication between the cells.

The **symplast** is the interconnected cytoplasm of plant cells, connected by plasmodesmata.

By the end of this section, you should be able to...

● describe the ultrastructure of the chloroplast, vacuole and tonoplast in eukaryotic cells, and relate these structures to their functions

Plant cells contain several kinds of organelle that are not found in animal cells. These include permanent vacuoles and chloroplasts.

Permanent vacuole

A vacuole is any fluid-filled space inside the cytoplasm surrounded by a membrane. Vacuoles occur quite frequently in animal cells, but they are only temporary, being formed and destroyed when needed. In non-woody plant cells the vacuole is a permanent structure with an important role. The vacuole can occupy up to 80% of the volume of a plant cell. It is surrounded by a specialised membrane called the **tonoplast**. The tonoplast has many different protein channels and carrier systems in it. It controls the movements of substances into and out of the vacuole and so controls the water potential of the cell. The vacuole is filled with **cell sap**, a solution of various substances in water. This solution causes water to move into the cell by **osmosis** (see **Section 4.1.3**), and as a result the cytoplasm is kept pressed against the cell wall. This in turn keeps the cells turgid (swollen) and the whole plant upright. The pressures that can be developed in this way are very high indeed. The pressure in a leaf cell can be up to 1500 kPa – in contrast, the pressure in a human artery when the heart is pumping blood out into the body is only 16 kPa.

As well as fulfilling the important role of maintaining the plant cell shape, the many different types of vacuoles in plants carry out a range of different functions. Vacuoles are used for the storage of a number of different substances. Many vacuoles store pigments; for example the betacyanin pigment of beetroot is normally stored in the vacuoles of the cells and does not leak out into the cytoplasm unless the root is cut. If the tissue is heated, the characteristics of the membrane around the vacuole will change and so pigment will leak out more rapidly. Vacuoles can store proteins in the cells of seeds and fruits, and in some plant cells they contain lytic enzymes and have a function rather like lysosomes in animal cells. Vacuoles often store waste products and other chemicals. For example, digitalis, a chemical found in foxgloves that can act both as a heart drug and a deadly poison, is stored in the vacuoles of the cells.

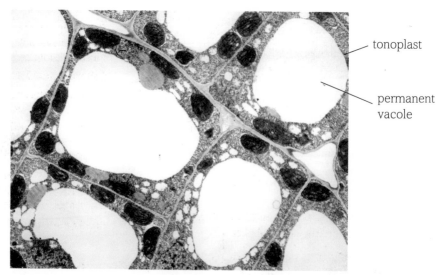

tonoplast

permanent vacole

fig A The tonoplast and the permanent vacuoles are key structures in the support systems of plants, but they have many other functions as well.

Chloroplasts

Of all the differences between plant and animal cells, the presence of **chloroplasts** in plant cells is probably the most important because they enable plants to make their own food. Not all plant cells contain chloroplasts – only those cells from the green parts of the plant. However, almost all plant cells contain the genetic information to make chloroplasts and so in some circumstances different areas of a plant will become green and start to photosynthesise. The exceptions are parasitic plants such as broomrape. Cells in flowers, seeds and roots contain no chloroplasts and neither do the internal cells of stems or the transport tissues. In fact the majority of plant cells do not have chloroplasts, but these organelles are very special and unique to plants.

(a)

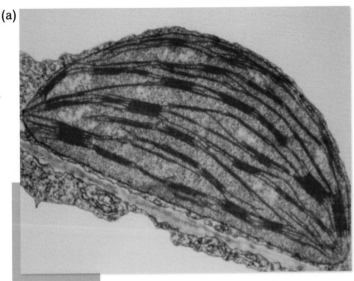

(b)

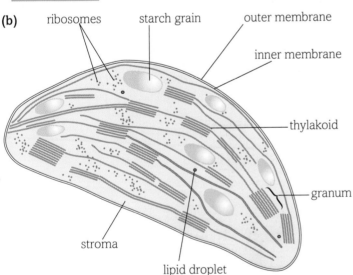

fig B (a) Micrograph of a chloroplast; and (b) Labelled diagram to show structures in a chloroplast.

There are some clear similarities between chloroplasts and mitochondria. Like mitochondria, chloroplasts:

- are large organelles: they have a biconvex shape with a diameter of 4–10 μm and are 2–3 μm thick

- contain their own DNA

- are surrounded by an outer membrane

- have an enormously folded inner membrane that gives a greatly increased surface area on which enzyme-controlled reactions take place

- are thought to have been free-living prokaryotic organisms that were engulfed by and became part of other cells at least 2000 million years ago.

However, there are also some clear differences. Chloroplasts:

- are the site of photosynthesis

- contain **chlorophyll**, the green pigment that is largely responsible for trapping the energy from light, making it available for the plant to use

- are formed from a type of relatively unspecialised plant organelle known as a leucoplast.

Amyloplasts

Amyloplasts are another specialised plant organelle and, like chloroplasts, they develop from leucoplasts. They are colourless and store starch (see **Section 1.2.2**). This can then be converted to glucose and used to provide energy when the cell needs it. Amyloplasts are found in large numbers in areas of a plant that store starch, for example potato tubers.

Questions

1 Amyloplasts and chloroplasts come from the same type of unspecialised cell. How do the two structures differ?

2 Compare and contrast the structure of a typical plant cell with the structure of a typical animal cell.

3 Explain why chloroplasts are found only in particular parts of a plant. Suggest what happens to make part of a plant, e.g. a potato tuber, turn green when exposed to light?

Key definitions

The **tonoplast** is the specialised membrane that surrounds the permanent vacuole in plant cells and controls movements of substances into and out of the cell sap.

Cell sap is the aqueous solution that fills the permanent vacuole.

Osmosis is a specialised form of diffusion that involves the movement of solvent molecules down a concentration gradient through a partially permeable membrane.

A **chloroplast** is an organelle adapted to carry out photosynthesis, containing the green pigment chlorophyll.

Chlorophyll is the green pigment that is largely responsible for trapping the energy from light, making it available for the plant to use in photosynthesis.

Amyloplasts are plant organelles that store starch.

By the end of this section, you should be able to...

● Explain how, in complex organisms, cells are organised into tissues, organs and organ systems

Multicellular organisms are made up of specialised cells but these cells do not operate on their own. The specialised cells are organised into groups of cells known as **tissues**. These tissues consist of one or more types of cells all carrying out a particular function in the body. However, tissues do not operate in isolation. Many tissues are further organised into **organs**.

Tissues

Tissues are groups of similar cells that all develop from the same kind of cell. Although there are many different types of specialised cells, there are only four main tissue types in the human body – epithelial tissue, connective tissue, muscle tissue and nervous tissue. Modified versions of these tissue types containing different specialised cells carry out all the functions of the body. **Fig A** shows some different **epithelial tissues**, which are tissues that form the lining of surfaces both inside and outside of the body. Although some epithelial tissues consist of more than one kind of cell, they all rest upon an extracellular basement membrane. Cells in epithelial tissues usually sit tightly together and form a smooth surface that protects the cells and tissues below.

Squamous epithelium is commonly found lining the surfaces of blood vessels, and forms the walls of capillaries and the lining of the alveoli. Cuboidal and columnar cells line many other tubes in the body. Ciliated epithelia often contain goblet cells that produce mucus. These epithelia form the surfaces of tubes in the gas exchange system and the oviducts. The regular waving of the cilia from side to side moves materials along inside the tubes. Compound epithelia are found where the surface is continually scratched and abraded, such as the skin. The thickness of the tissue protects what lies beneath as new cells continue to grow.

There are many other tissues in the body, including muscle tissue, nervous tissue, the collagen tissue and elastin tissue found in artery walls and the glandular tissue that secretes substances from inside the cells. Connective tissue is the main supporting tissue in the body, and includes bone tissue and cartilage tissue as well as packing tissue that supports and protects some of the organs. Some of these tissues are shown in **fig B**.

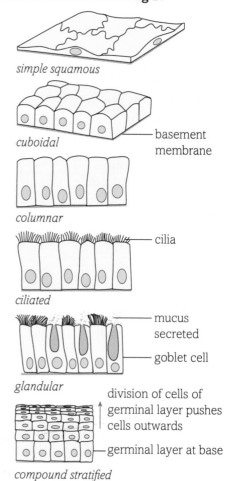

simple squamous

cuboidal — basement membrane

columnar

ciliated — cilia

glandular — mucus secreted — goblet cell

compound stratified — division of cells of germinal layer pushes cells outwards — germinal layer at base

fig A There are many kinds of epithelial tissues inside the human body.

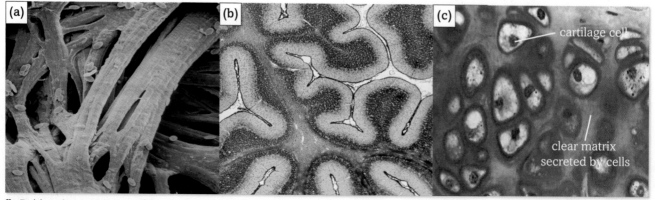

fig B (a) cardiac muscle tissue; (b) brain tissue; (c) cartilage tissue.

(c) — cartilage cell — clear matrix secreted by cells

Organs

An organ is a structure made up of several different tissues that work effectively together to carry out a particular function. There are many organs in the human body, some of which are shown in **fig C**. Plants also have cells grouped into tissues and organs. For example the leaf is an organ that is composed of vascular tissue, epithelial tissue and mesophyll tissues as shown in **fig D**.

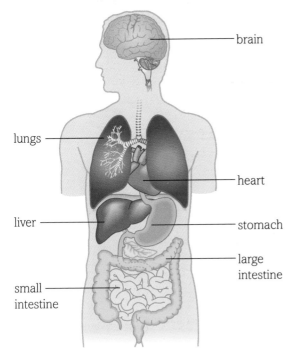

fig C Some of the organs and organ systems of the human body.

Systems

In animals, in many cases a number of organs work together as an **organ system** to carry out large-scale functions in the body. For example the digestive system includes the organs of the stomach, pancreas, small and large intestines, and the nervous system includes the brain, spinal cord and all peripheral nerves.

Most of the cells in tissues, organs and systems have differentiated during development so that they are capable of carrying out their specific function. You will find out more about how this process happens in **Book 2 Chapter 7.2**.

Questions

1 Explain how the structure of the following tissues is related to their function.
   (a) squamous epithelium lining an alveolus
   (b) ciliated epithelium lining a bronchus
   (c) muscle tissue in the biceps muscle.

2 (a) Choose one of the systems in the human body and describe briefly the cells, tissues and organs found within that system.
   (b) Explain why this grouping enables the system to carry out its function effectively.

Key definitions

A **tissue** is a group of specialised cells carrying out a particular function in the body.

An **organ** is a structure made up of several different types of tissues grouped together to carry out a particular function in the body.

Epithelial tissues are tissues that form the lining of surfaces inside and outside the body.

An **organ system** is a group of organs working together to carry out particular functions in the body.

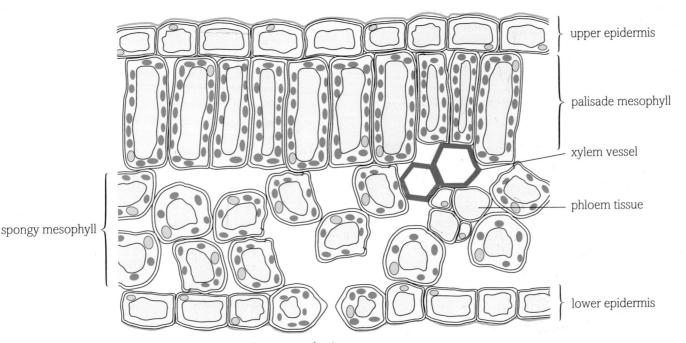

fig D Some of the tissues found in a leaf – the photosynthetic organ of a plant.

1 The photograph below shows a mitochondrion as seen using an electron microscope.

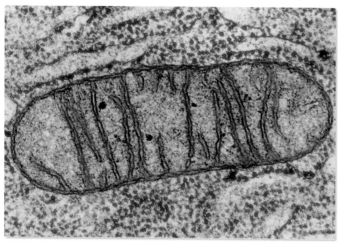

(a) Describe the role of mitochondria in a cell. [2]
(b) Make an accurate drawing of this mitochondrion enlarged ×2. On your drawing label the **matrix** and a **crista**. [4]

[Total: 6]

2 The photograph below shows a chloroplast as seen using an electron microscope.

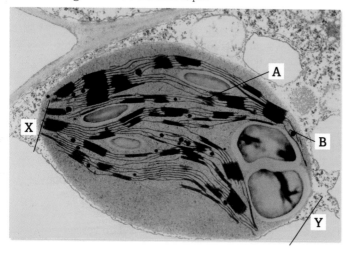

(a) Name the parts labelled **A** and **B**. [2]
(b) The actual length of the chloroplast between **X** and **Y** is 5 µm. Calculate the magnification of this chloroplast. Show your working. [3]
(c) Name **one** type of cell that contains chloroplasts. [1]

[Total: 6]

3 (a) Draw and label a diagram to show the structure of a chloroplast, as seen using an electron microscope. [4]
(b) The photograph below shows a group of mitochondria in a liver cell, as seen using an electron microscope. The magnification is ×50 000.

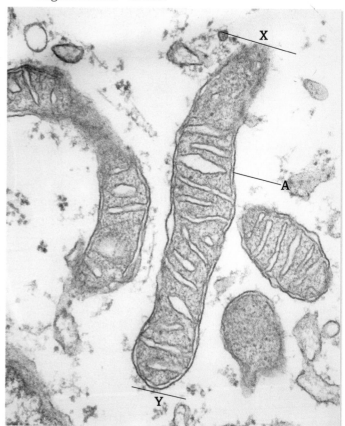

(i) Measure the length of the mitochondrion labelled **A** between **X** and **Y**. Calculate the actual length of this mitochondrion in µm. Show your working. [3]
(ii) Name **one** other structure that might be visible in the cytoplasm of this liver cell if the magnification used was higher. [1]
(iii) Give one reason why the double membrane is not clearly visible all around the mitochondrion labelled **A**. [1]

[Total: 9]

4 The table below refers to some cell structures. Complete the table by inserting the correct word, words, or diagram in the appropriate boxes. Leave the shaded grey boxes empty.

Name of cell structure	Description of cell structure	Diagram of cell structure
[1]	1 Darkly-stained region in the nucleus. 2 Where ribosomal RNA is made.	
Centrioles		[2]
Lysosome	1 2 [2]	
[1]	1 Hollow cylinders made of protein. 2 Form spindle fibres.	

[Total: 6]

5 The photograph below shows some onion cells as seen using the high power of a light microscope.

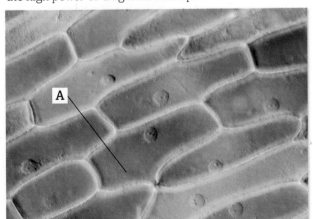

(a) Make an accurate drawing, enlarged ×2, of the cell labelled **A**.
    Do **not** label your drawing.   [3]

(b) All the onion cells have a cell surface (plasma) membrane. The diagram below shows the structure of this membrane.

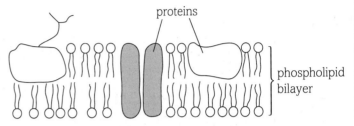

Explain how the properties of phospholipids result in the formation of a bilayer.   [3]

[Total: 6]

6 The endosymbiotic theory received fresh impetus from a 1967 paper by Lynne Margulis, who offered evidence from microbiology.
   (a) What is a symbiotic relationship?   [1]
   (b) What do chloroplasts and mitochondria have in common with prokaryotic organisms?   [1]
   (c) Both chloroplasts and mitochondria have double membranes. When a white blood cell engulfs bacteria it forms a vacuole around them. Describe how this adds support to the endosymbiotic theory.   [1]
   (d) Further work by Lynn Margulis in 1981 argued that eukaryotic flagella came from bacteria called spirochaetes. Explain why this has not received much support.   [1]

[Total: 4]

7 Scientists using light microscopes were unable to distinguish small organelles like ribosomes.
   (a) Name **three** other organelles that were too small for these scientists to see.   [3]
   (b) Name **two** organelles that would have been seen using the light microscope.   [2]
   (c) Light microscopes can be used to watch substances move in real time. Explain why this is impossible in an electron microscope.   [2]

[Total: 7]

8 Outline the stages of the production of the primary and secondary cell walls in a perennial plant.   [6]

[Total: 6]

CHAPTER 2.2 Prokaryotic cells

Introduction

The human body contains around 10 times more bacterial cells than it does human cells – on a numbers basis, we are more prokaryotic than eukaryotic! Bacteria are found on our skin, in our nose and most of all in our digestive systems. But bacteria are so small that even the trillions found in the human body have an estimated mass of only between 0.9 and 2.7 kg, which is just 1–3% of the total body mass. In the Human Microbiome Project, scientists have sequenced the genomes of bacteria taken from healthy humans. They have identified around 10 000 species of bacteria in the human ecosystem alone. Many of these bacteria have a direct and positive effect on our health. For example, bacteria help us digest food and absorb nutrients in the digestive system.

The ultrastructure of prokaryotic cells differs from that of eukaryotic cells in a number of fundamental ways, and in this chapter you will be looking at the similarities and differences between these groups of organisms. Prokaryotes do not have membrane-bound organelles so the genetic material is free in the cytoplasm. They may have extra bits of genetic material called plasmids and the ribosomes are different in their chemical make-up. Bacterial cell walls are unique – and they vary considerably. You will discover how stains can be used to identify some of these differences, which are important in pathogenic bacteria.

You will also look at viruses. These are the ultimate parasites, taking over the genetic material of other organisms to replicate and make more viruses. Most naturally occurring viruses cause disease, so it is important to understand as much as we can about their structures and life cycles to help defend ourselves against them.

All the maths you need

- Make use of appropriate units (*e.g. relative sizes of eukaryotic cells, prokaryotic cells and viruses*)
- Carry out calculations using numbers in standard and ordinary form (*e.g. use of magnification*)
- Make order of magnitude calculations (*e.g. use and manipulate the magnification formula:*
$$magnification = \frac{size\ of\ image}{size\ of\ real\ object})$$
- Use and manipulate equations, including changing the subject of an equation (*e.g. magnification*)

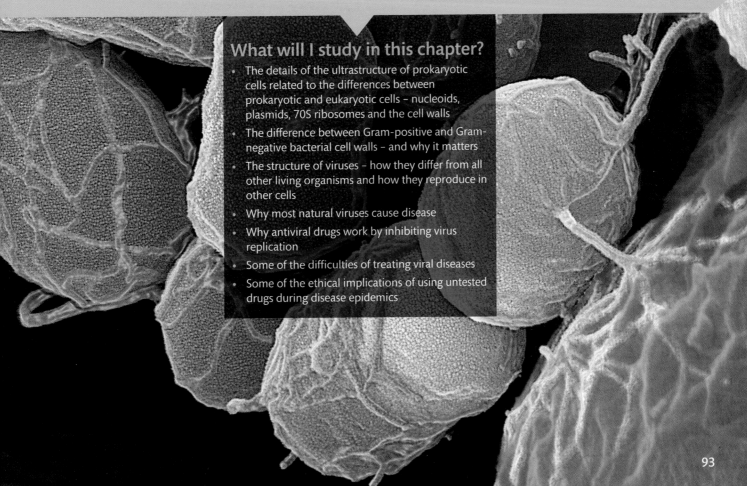

What will I study later?

- Classification of prokaryotes using the evidence of DNA sequencing
- How to culture bacteria in an aseptic environment on different media
- Different methods of measuring the growth of bacterial cultures
- The phases of the growth of a bacterial colony (bacterial growth curves)
- Bacteria as pathogens, including some of the diseases they cause and how they can be treated or prevented (A level)
- The development and spread of antibiotic resistance in bacteria (A level)
- How the influenza virus causes disease (A level)
- Ways of controlling endemic diseases (A level)
- How the immune system recognises pathogens such as bacteria and viruses and destroys them (A level)
- The use of bacterial plasmids and viruses in gene technology (A level)
- The significance of bacteria in recycling nutrients within ecosystems (A level)

What have I studied before?

- How the main sub-cellular structures of prokaryotic cells are related to their functions, including the genetic material, plasmids and cell membranes
- How communicable diseases are caused by bacteria and viruses (as well as fungi and protoctists)

What will I study in this chapter?

- The details of the ultrastructure of prokaryotic cells related to the differences between prokaryotic and eukaryotic cells – nucleoids, plasmids, 70S ribosomes and the cell walls
- The difference between Gram-positive and Gram-negative bacterial cell walls – and why it matters
- The structure of viruses – how they differ from all other living organisms and how they reproduce in other cells
- Why most natural viruses cause disease
- Why antiviral drugs work by inhibiting virus replication
- Some of the difficulties of treating viral diseases
- Some of the ethical implications of using untested drugs during disease epidemics

By the end of this section, you should be able to...

● describe the ultrastructure of prokaryotic cells and their organelles including nucleoid, plasmids, 70S ribosomes and the cell wall

● distinguish between Gram-positive and Gram-negative bacterial cell walls and explain why each type responds differently to some antibiotics

Bacteria, cyanobacteria and archaebacteria are prokaryotic organisms. Bacteria alone are probably the most common form of life on Earth. Some bacteria are pathogens and cause disease, but the great majority do no harm and many are beneficial to living organisms, for example as gut bacteria and in the cycling of nutrients in the natural world. (see **Book 2 Chapter 10.2**). In this section, you will mainly consider the structure and functions of bacterial cells.

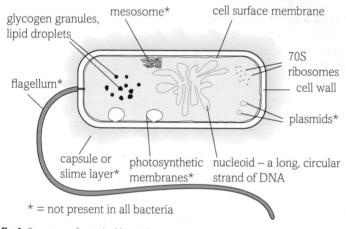

fig A Structure of a typical bacterium.

The structure of bacteria

All bacterial cells have certain features in common, although these vary greatly between species.

Bacterial cell walls

All bacterial cells have a cell wall. The contents of bacterial cells are usually **hypertonic** to the medium around them, so water tends to move into the cells by osmosis. The cell wall prevents the cell swelling and bursting. It also maintains the shape of the bacterium, and gives support and protection to the contents of the cell. All bacterial cell walls consist of a layer of **peptidoglycan** that is made up of many parallel polysaccharide chains with short peptide cross-linkages forming an enormous molecule with a net-like structure. Some bacteria have a capsule (or slime layer if it is very thin and diffuse) around their cell walls. This may be formed from starch, gelatin, protein or glycolipid, and protects the bacterium from phagocytosis by white blood cells. It also covers the cell markers on the cell membrane that identify the cell. So a capsule can make it easier for a bacterium to be pathogenic (to cause disease) because it is not so easily identified by the immune system. This is the case for the bacteria that cause pneumonia, meningitis, tuberculosis (TB) and septicaemia. However, many capsulated bacteria do not cause disease. It seems likely that capsules evolved to help the bacteria survive very dry conditions.

Pili and flagellae

Some bacteria have from one to several hundred thread-like protein projections from their surface. These are called the **pili** (or **fimbriae**) and they are found on some well-known bacteria such as *Escherichia coli* (*E. coli*) and *Salmonella* spp. They seem to be used for attachment to a host cell and for sexual reproduction. However, they also make bacteria more vulnerable to virus infections, as a **bacteriophage** can use pili as an entry point to the cell.

Some bacteria can move themselves using **flagella**. These are little bigger than one of the microtubules contained in a eukaryotic flagellum, and are made of a many-stranded helix of the protein flagellin. The flagellum moves the bacterium by rapid rotations – about 100 revolutions per second.

Cell surface membrane

The cell surface membrane in prokaryotes is similar in both structure and function to the membranes of eukaryotic cells. However, bacteria have no mitochondria so the cell membrane is also the site of some of the respiratory enzymes. In some bacterial cells such as *Bacillus subtilis*, a common soil bacterium, the membrane shows infoldings known as **mesosomes**. There is still some debate about their function. Some scientists think they may be an artefact from the process of preparing the cell for an electron micrograph, others believe they are associated with enzyme activity, particularly during the separation of DNA and the formation of new cross walls during replication. It appears that other infoldings of the bacterial cell surface membrane may be used for photosynthesis by some bacterial species.

Plasmid

Some bacterial cells also contain one or more much smaller circles of DNA known as **plasmids**. A plasmid codes for a particular aspect of the bacterial phenotype in addition to the genetic information in the nucleoid, for example the production of a particular toxin or resistance to a particular antibiotic. Plasmids can reproduce themselves independently of the nucleoid. They can be transferred from one bacterium to another in a form of sexual reproduction using the pili.

Nucleoid

The genetic material of prokaryotic cells consists of a single length of DNA, often circular, which is not contained in a membrane-bound nucleus. However, the DNA is folded and coiled to fit into the bacterium. The area in the bacterial cell where this DNA tangle is found is known as the **nucleoid**. In an *E. coli* bacterium it takes up about half of the area of the bacterium.

70S ribosomes

The bacteria, cyanobacteria and archaebacteria have no membrane-bound organelles, but they do have ribosomes where they carry out protein synthesis. The ribosomes in bacterial cells are 70S, smaller than the 80S ribosomes in eukaryotes. They have two subunits. The smaller is 30S and the larger is 50S (see **Section 2.1.4**). They are involved in the synthesis of proteins in a similar way to eukaryotic ribosomes.

Gram staining and bacterial cell walls

Whilst all bacterial cell walls contain peptidoglycan, there are in fact two main types of bacterial cell wall. These can be distinguished by **Gram staining**, a staining technique developed by Christian Gram (1853–1938) in 1884 and still in use today. It is valuable because different types of bacteria are vulnerable to different types of antibiotics and one of the factors that affects their vulnerability is the type of cell wall.

Before staining, bacteria are often colourless. The cell walls of **Gram-positive bacteria** (e.g. methicillin-resistant *Staphylococcus aureus*, MRSA) have a thick layer of peptidoglycan containing chemicals such as **teichoic acid** within its net-like structure. The crystal violet/iodine complex in the Gram stain is trapped in the thick peptidoglycan layer and resists decolouring when the bacteria are dehydrated using alcohol. As a result it does not pick up the red safranin counterstain, leaving the positive purple/blue colour.

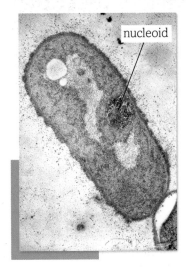

fig B The nucleiod area of a bacterium.

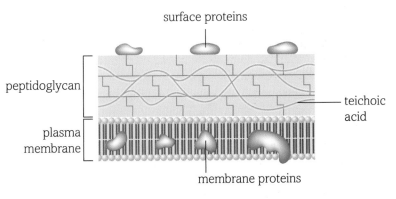

Gram-positive bacterial cell walls

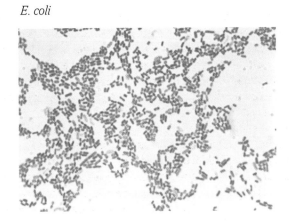

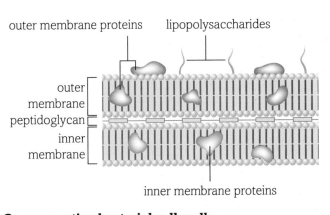

Gram-negative bacterial cell walls

fig C The difference in the cell wall structure of the bacteria results in the different reactions with the Gram stain.

The cell walls of **Gram-negative bacteria** have a thin layer of peptidoglycan with no teichoic acid between two layers of membrane. The outer membrane is made up of lipopolysaccharides. After the crystal violet/iodine complex is applied, the bacteria are dehydrated in ethanol. The lipopolysaccharide layer dissolves in the ethanol leaving the thin peptidoglycan layer exposed. The crystal violet/iodine complex is washed out and the peptidoglycan takes up the red safranin counterstain, so the cells appear red.

Antibiotics and bacterial cell walls

Antibiotics are drugs that are used against bacterial pathogens. There are a number of different types of antibiotics, each working in different ways. They may work by affecting the bacterial cell walls, the cell membranes, the genetic material, the enzymes or the ribosomes. Antibiotics usually target features of bacterial cells that differ from those of eukaryotic cells, including the bacterial cell walls and the 70S ribosomes.

Different types of bacteria are sensitive to different types of antibiotics. Doctors need to know if a pathogenic bacterium is Gram-positive or Gram-negative as this will affect the choice of antibiotic used to treat the disease.

To pinpoint the actions of an antibiotic, first think about the difference between human cells and bacterial cells, and then about the differences between Gram-positive and Gram-negative bacteria.

Some antibiotics, such as beta-lactam antibiotics (penicillins and cephalosporins), inhibit the formation of the peptidoglycan layer of the bacterial cell wall. As a result they are very effective against Gram-positive bacteria, as they have a thick peptidoglycan layer on the surface of the cell, but less effective against Gram-negative bacteria, as their peptidoglycan layer is hidden and less vital to the wall structure. They don't affect human cells as they don't have a peptidogycan cell wall at all.

Glycopeptide antibiotics, such as vancomycin, are large polar molecules that cannot penetrate the outer membrane layer of Gram-negative bacteria. However, they are very effective against Gram-positive bacteria, even ones that have developed resistance to many other antibiotics.

Polypeptide antibiotics, such as polymixins, are rarely used as they can have serious side effects. They are very effective against Gram-negative bacteria because they interact with the phospholipids of the outer membrane. They do not affect Gram-positive bacteria.

Most other antibiotics affect both Gram-positive and Gram-negative bacteria because they target common processes such as protein synthesis by the ribosomes. They only target prokaryote ribosomes, not eukaryotic ones.

If you are studying A level Biology, you will learn more about antibiotics in Book 2.

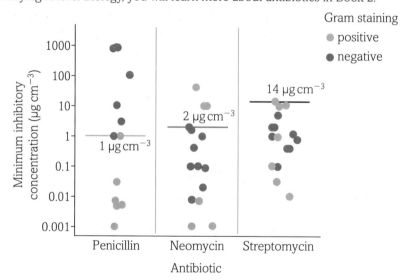

fig D This graph shows that penicillin is effective against Gram-positive bacteria as all the blue circles are on or below 1μg cm⁻³, so all types of Gram-positive bacteria are killed at a relatively low dose. Neomycin is best for Gram-negative bacteria as all the red circles are below 2 μg cm⁻³. Streptomycin is the antibiotic to choose if the Gram status of the bacteria is unknown because it kills all types of bacteria at a dose of only 14 μg cm⁻³.

Alternative ways of classifying bacteria

Grouping bacteria simply by the way their cell walls do or do not take up Gram stains is of limited use in classifying the different types. Another way in which bacteria can be identified is by their shape. Some bacteria are spherical (**cocci**) while the **bacilli** are rod-shaped. Yet others are twisted (**spirilla**) or comma-shaped (**vibrios**).

Bacteria are also sometimes grouped by their respiratory requirements. **Obligate aerobes** need oxygen for respiration. **Facultative anaerobes** use oxygen if it is available, but can manage without it. Many human pathogens fall into this group. **Obligate anaerobes** can only respire in the absence of oxygen – in fact oxygen will kill them.

Questions

1   Make a table to compare and contrast prokaryotic and eukaryotic cells.

2   What is the difference in the structure of the walls of Gram-positive and Gram-negative bacteria?

3   (a) How does the structure of the walls of Gram-positive and Gram-negative bacteria affect the effectiveness of some antibiotics?

   (b) Using the data in **fig D** suggest why streptomycin would be the best of these three antibiotics to use if you did not know whether a bacterial pathogen was Gram-positive or Gram-negative.

Key definitions

A **hypertonic solution** is a solution with a higher concentration of solutes and lower concentration of water (solvent) than the surrounding solution.

Peptidoglycan is a large, net-like molecule found in all bacterial cell walls made up of many parallel polysaccharide chains with short peptide cross-linkages.

Pili (fimbriae) are thread-like protein projections found on the surface of some bacteria.

Bacteriophages are viruses that attack bacteria.

Flagella are many-stranded helices of the contractile protein flagellin found on some bacteria. They move the bacteria by rapid rotations.

Mesosomes are infoldings of the cell membrane of bacteria.

A **nucleoid** is the area in a bacterium where we find the single length of coiled DNA.

Plasmids are small, circular pieces of DNA that code for specific aspects of the bacterial phenotype.

Gram staining is a staining technique used to distinguish types of bacteria by their cell wall.

Gram-positive bacteria are bacteria that contain teichoic acid in their cell walls and stain purple/blue with Gram staining.

Teichoic acid is a chemical found in the cell walls of Gram-positive bacteria.

Gram-negative bacteria are bacteria that have no teichoic acid in their cell walls. They stain red with Gram staining.

Cocci are spherical bacteria.

Bacilli are rod-shaped bacteria.

Spirilla are bacteria with a twisted or spiral shape.

Vibrios are comma-shaped bacteria.

Obligate aerobes are organisms that need oxygen for respiration.

Facultative anaerobes are organisms that use oxygen if it is available, but can respire and survive without it.

Obligate anaerobes are organisms that can only respire in the absence of oxygen and are killed by oxygen.

By the end of this section, you should be able to...

● recognise that viruses are not living cells

● explain the classification of viruses based on structure and nucleic acid types as illustrated by λ (lambda) phage (DNA), tobacco mosaic virus and Ebola (RNA) and human immunodeficiency virus (RNA retrovirus)

● describe the lytic cycle of a virus and explain latency

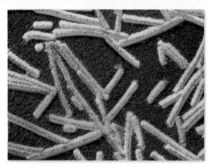

fig A The tiny rod-shaped particles of the tobacco mosaic virus, seen here under the scanning electron microscope, can cause serious damage to a crop.

Viruses are the smallest of all the microorganisms, and range in size from 0.02 μm to 0.3 μm across, about 50 times smaller than the average bacterium. Viruses are not cells. They are arrangements of genetic material and protein that invade other living cells and take over their biochemistry to make more viruses. It is because of this reproduction and the fact that they change and evolve in an adaptive way, that they are classed as living organisms.

Viruses

Most scientists working on viruses class them as obligate intracellular parasites, meaning they can only exist and reproduce as parasites in the cells of other living organisms. Because natural viruses invade and take over living cells to reproduce, they usually all cause damage and disease of some sort. They can withstand drying and long periods of storage whilst maintaining their ability to infect cells. There are very few drugs that have any effect on viruses, and those that do only work in very specific instances; for example, acyclovir can help prevent herpes (cold sores) and genital herpes.

Did you know?

Discovering viruses

People suspected the presence of viruses causing disease in the late nineteenth century. They were developed as a model to explain the way certain diseases were passed from one individual to another, but it was not until 1935 that the first virus was identified by Wendell Stanley (1904–71).

The leaves of tobacco plants are prone to an unpleasant blotchy disease that has a devastating effect on the plants, and no-one could find the cause. Stanley pressed the juice from around 1300 kg of diseased tobacco leaves. After extraction and purification, this produced pure, needle-like crystals which, if dissolved in water and painted onto tobacco leaves, caused the symptoms of the disease. The particles were called tobacco mosaic virus (TMV). It was obvious that the crystals were not living in the usual sense of the word, yet they retained the ability to cause disease. Viruses cannot be seen using a light microscope because they are usually smaller than half a wavelength of light. With the development of the electron microscope the TMV particles were shown to be rod-like structures with a protein coat formed around a core of RNA.

The structure of viruses

Viruses usually have geometric shapes and similar basic structures. However, there is considerable variation in the genetic material they possess, the structure of their protein coat and whether or not they have an **envelope**. The protein coat or **capsid** is made up of simple repeating protein units known as **capsomeres**, arranged in different ways. Using repeating units minimises the amount of genetic material needed to code for coat production. It also makes sure that assembling the protein coat in the host cell is as simple as possible. In some viruses the genetic material and protein coat is covered by a lipid envelope, produced from the host cell. The presence of the envelope makes it easier for the viruses to pass from cell to cell, but it does make them vulnerable to substances such as ether, which will dissolve the lipid membrane.

Classifying viruses

Viruses attach to their host cells by means of specific proteins (antigens) known as **virus attachment particles (VAPs)** that target proteins in the host cell surface membrane. Because they respond to particular molecules of the host cell surface, viruses are often quite specific in the tissue they attack.

Viruses are classified by their genome and their mode of replication. Viral genetic material can be DNA or RNA, and the nucleic acid is sometimes double-stranded and sometimes single. The way in which the viral genetic material is used in the host cell to make new viruses depends on which form it is in:

● **DNA viruses**: In these viruses the genetic material is DNA. The viral DNA acts directly as a template for both new viral DNA and for the mRNAs needed to induce synthesis of viral proteins. Examples of DNA viruses include the smallpox virus, adenoviruses which cause colds, and some bacteriophages (viruses which infect bacteria) for example the λ (lambda) phage in **fig B**.

● **RNA viruses**: 70% of viruses have RNA as their genetic material and they are much more likely to mutate than DNA viruses. RNA viruses do not produce DNA as part of their life cycle. The majority of RNA viruses contain a single strand of RNA and are know as ssRNA viruses. Positive ssRNA viruses (also known as positive-sense ssRNA viruses) have RNA that

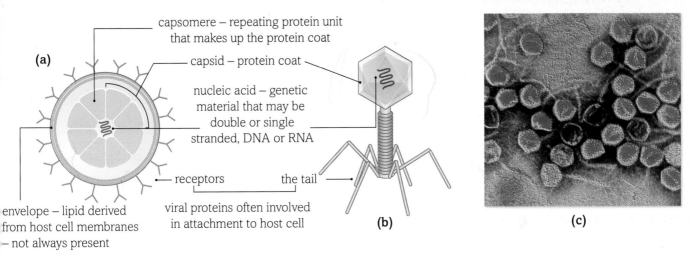

fig B General viral structures: (a) RNA virus; (b) λ (lambda) phage; (c) electron micrograph of λ (lambda) bacteriophages.

can act directly as mRNA and be translated at the ribosomes. Examples of plant and animal diseases caused by positive ssRNA viruses include tobacco mosaic viruses, SARS, polio and hepatitis C. Negative ssRNA viruses (also known as negative-sense ssRNA viruses) cannot be directly translated. The RNA strand must be transcribed before it is translated at the ribosomes. Examples of diseases caused by negative ssRNA viruses include measles, influenza and Ebola.

- **RNA retroviruses**: Retroviruses are a special type of RNA virus. They have a protein capsid and a lipid envelope. The single strand of viral RNA directs the synthesis of a special enzyme called **reverse transcriptase**. This goes on to make DNA molecules corresponding to the viral genome. This DNA is then incorporated into the host cell DNA and used as a template for new viral proteins and ultimately a new viral RNA genome. HIV (human immunodeficiency virus) is a retrovirus and some forms of leukaemia are also caused by this type of virus.

How viruses reproduce

Natural viruses all cause disease, and they attack every other known type of living organism. There are even viruses that attack bacteria, known as bacteriophages. We are constantly involved in a battle against the viruses that cause disease in ourselves, our animals, our crops and our environment. In order to understand how viruses cause damage to the body, and to be able to try to target drugs effectively, it is important to understand how they reproduce in the human body.

Virus 'life cycles'

Viruses only reproduce within the cells of the body. They attack their host cells in a number of different ways. For example, bacteriophages inject their genome into the host cell, but the bulk of the viral material remains outside the bacterium. The viral DNA forms a circle or plasmid within the bacterium.

The viruses that infect animals get into the cells in several ways. Some types are taken into the cell by endocytosis – either with or without the envelope – and the host cell then digests the capsid, releasing the viral genetic material. Most commonly, the viral envelope fuses with the host cell surface, releasing the rest of the

virus into the cell membrane. Plant viruses usually get into the plant cell using a vector, often an insect such as an aphid, to pierce the cell wall.

DNA virus replication

Once a virus is in the host cell there are two different routes of infection:

Latency – the lysogenic pathway

Many DNA viruses are **non-virulent** when they first get into the host cell. They insert their DNA into the host DNA so it is replicated every time the host cell divides. This DNA inserted into the host is called a **provirus**. Messenger RNA is not produced from the viral DNA because one of the viral genes causes the production of a repressor protein that makes it impossible to translate the rest of the viral genetic material. The virus does not affect the host cell or make the host organism ill at this stage in the life cycle. During this period of **lysogeny**, when the virus is part of the reproducing host cells, the virus is said to be **latent**.

The lytic pathway

Sometimes the viral genetic material is replicated independently of the host DNA straight after entering the host. Mature viruses are made and eventually the host cell bursts, releasing large numbers of new virus particles to invade other cells. The virus is said to be **virulent** (disease causing) and the process of replicating and killing cells is known as the lytic pathway. Under certain conditions, such as when the host is damaged, viruses in the lysogenic state are activated. The amount of repressor protein decreases and the viruses enter the lytic pathway and become virulent (see **fig C**).

Some types of virus have both latent and lytic stages in their life cycle, but others move straight to the lytic stage after they have infected a cell.

RNA virus replication

There are a number of different types of RNA viruses and they replicate themselves in different ways.

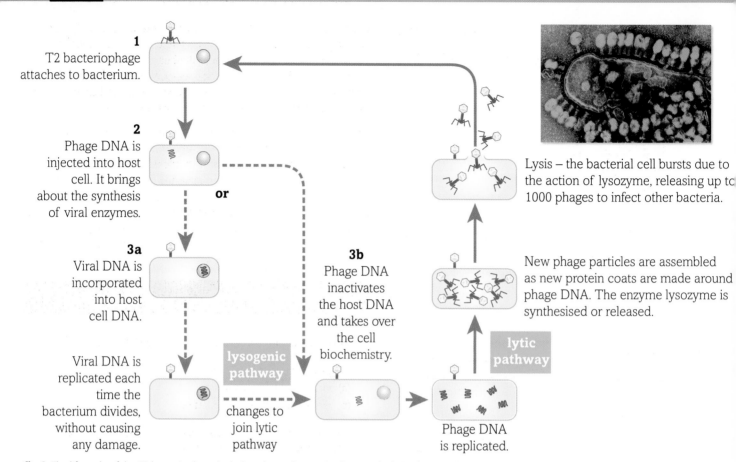

1
T2 bacteriophage attaches to bacterium.

2
Phage DNA is injected into host cell. It brings about the synthesis of viral enzymes.

or

3a
Viral DNA is incorporated into host cell DNA.

Viral DNA is replicated each time the bacterium divides, without causing any damage.

changes to join lytic pathway

lysogenic pathway

3b
Phage DNA inactivates the host DNA and takes over the cell biochemistry.

Phage DNA is replicated.

lytic pathway

New phage particles are assembled as new protein coats are made around phage DNA. The enzyme lysozyme is synthesised or released.

Lysis – the bacterial cell bursts due to the action of lysozyme, releasing up to 1000 phages to infect other bacteria.

fig C The life cycle of the T2 bacteriophage includes a latent, lysogenic phase and a lytic phase.

Positive ssRNA viruses

These are viruses that contain a single strand of RNA that is a sense strand. It is used directly as mRNA for translation into proteins at the ribosomes. The proteins made include viral structural proteins and an RNA polymerase, which is used to replicate the viral RNA.

Negative ssRNA viruses

The single strand of RNA in these viruses is an antisense strand. Before it can be used to make viral proteins and more viral RNA it must be transcribed into a sense strand. The virus imports RNA replicase, which uses free bases in the host cell to transcribe the antisense RNA strand and produce a sense strand that can be translated at the ribosomes. Once the RNA strand has been transcribed it acts as mRNA at the ribosomes and codes for viral proteins including RNA replicase. These viral proteins combine with replicated viral RNA to form new viral particles.

RNA retroviruses

Retroviruses, including the HIV virus that causes AIDS and the Rous sarcoma virus that causes cancer in chickens, have a rather different and complex life cycle. They have viral RNA as their genetic material. It cannot be used as mRNA, but is translated into DNA by the viral enzyme reverse transcriptase in the cytoplasm of the cell. This viral DNA passes into the nucleus of the host cell where it is inserted into the host DNA. Host transcriptase enzymes then make viral mRNA and new viral genome RNA. New viral

material is synthesised and the new viral particles leave the cell by exocytosis (see **Section 2.1.4**). The host cell continues to function as a virus-making factory, while the new viruses move on to infect other cells.

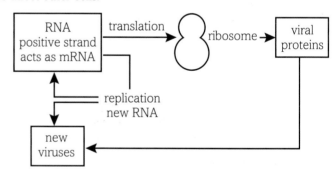

fig D Replication of a positive ssRNA virus.

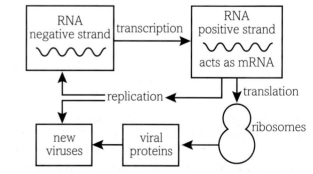

fig E Replication of a negative ssRNA virus.

1 The retrovirus attacks an animal cell.

retrovirus (HIV)

animal cell

2 Viral RNA enters the host cell. This RNA cannot be used as mRNA.

3 Viral RNA is translated into viral DNA by reverse transcriptase in the cytoplasm.

viral reverse transcriptase

viral DNA

4 Viral DNA is incorporated into the host DNA in the nucleus. It directs the production of new viral genome RNA, mRNA and coat proteins.

host DNA

viral DNA

5 New viral particles are assembled and leave the host cell by exocytosis. Viral DNA remains in the nucleus so the process is repeated.

New virus particles escape by exocytosis.

The cell continues to act as a virus factory.

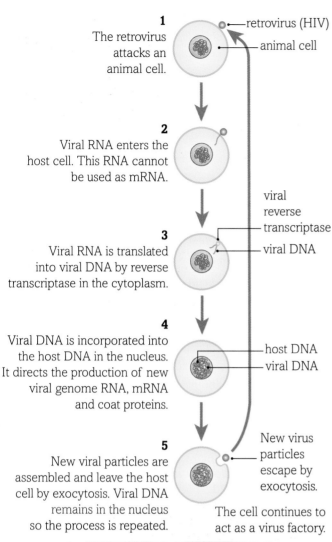

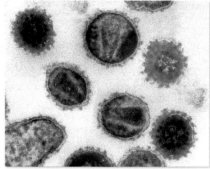

fig F The life cycle of a retrovirus.

Viruses and disease

Viruses cause disease in animals, plants and even in bacteria. They can cause the symptoms of disease by the lysis of the host cells, by causing the host cells to release their own lysosomes (see **Section 2.1.4**) and digest themselves from the inside or by the production of toxins that inhibit cell metabolism.

Viral infections are often specific to particular tissues. For example, adenoviruses, which cause colds, affect the tissues of the respiratory tract, but do not damage the cells of the brain or the intestine. This specificity seems to be due to the presence or absence of cell markers on the surface of host cells. Each type of cell has its own recognition markers and different types of virus can only bind to particular markers. The presence or absence of these markers can even affect whether a group of living organisms is vulnerable to attack by viruses at all. For example, the angiosperms (flowering plants) are vulnerable to viral diseases, but the gymnosperms (conifers and their relatives) are not.

Viruses are well-known for causing diseases like flu, measles, AIDS and Ebola. Research also shows that in some cases they play a role in the development of cancers. Certain animal cancers have been clearly linked to viral infection, and in humans there seems to be a link in certain specific cases. For example, the human papilloma virus responsible for warts on the skin, including genital warts, has been linked with the occurrence of pre-cancerous and cancerous changes in the cells of the cervix, and there is now a vaccine against it.

Questions

1 What adaptations make viruses such successful pathogens?

2 Suggest valid arguments for the case that:
   (a) viruses are living organisms
   (b) viruses are **not** living organisms.

3 What are the main differences between the lytic and lysogenic pathways of infection by DNA viruses?

4 Make a table to compare the different ways in which RNA viruses reproduce.

Key definitions

An **envelope** is a coat around the outside of a virus derived from lipids in the host cell.

The **capsid** is the protein coat of a virus.

Capsomeres are the repeating protein units that make up the capsid of a virus.

Virus attachment particles (VAPs) are specific proteins (antigens) that target proteins in the host cell surface membrane.

DNA viruses are composed of DNA as the genetic material.

RNA viruses are composed of RNA as the genetic material.

Retroviruses are a special type of RNA virus that control the production of DNA corresponding to the viral RNA and insert it into the host cell DNA.

Reverse transcriptase is an enzyme synthesised in the life cycle of a retrovirus that makes DNA molecules corresponding to the viral RNA genome.

Non-virulent is a term used to describe a microorganism that is not disease-causing.

A **provirus** is the DNA that is inserted into the host cell during the lysogenic pathway of reproduction in viruses.

Lysogeny is the period when a virus is part of the reproducing host cell, but does not affect it adversely.

Latent is the state of the non-virulent virus within the host cell.

Virulent is a term used to describe a microorganism that is disease-causing.

By the end of this section, you should be able to...

- Describe how antiviral medicines work by inhibiting viral replication because viruses are not living cells
- Explain how control of viral infections focuses on the prevention of the spread of disease as viral infections are difficult to treat
- Evaluate the ethical implications of using untested drugs during epidemics

As you have seen, the lifecycle of a virus involves the destruction of host cells. As a result of this direct damage, and the response of the host body to infection, viruses usually cause disease in the organisms they infect.

The spread of viral diseases

Viral diseases are spread in many different ways. The key feature is that material carrying viruses from an infected animal or plant comes into contact with vulnerable tissues in another uninfected organism. So, viruses may be spread through infected mucus, droplets of saliva, infected blood or faeces, or simple contact between infected organisms. International travel means that diseases that would once have just caused local outbreaks can now rapidly spread all over the world. Different viral diseases are spread in different ways. For example:

- Foot-and-mouth disease is a serious disease of cloven-hoofed animals such as cattle. It severely weakens adult animals and kills a high percentage of young animals. It is spread through body secretions, such as milk and semen, and transmitted in the breath and the faeces of infected animals. Healthy animals can pick up the virus from contaminated pens, food, water, contact with diseased animals and even from infected meat and animal products if they are eaten.
- **Ebola** is a severe viral illness caused by the Ebola virus. It is often fatal, especially if the symptoms are untreated. It is an animal disease that spreads to humans through the faeces, urine, blood and meat of infected animals. It then spreads easily from person to person by the direct contact of the skin or mucous membranes of a healthy person with blood, faeces and other body secretions of an infected person, or even bedding and surfaces contaminated with fluids from an infected person.

Treating viral diseases

As you have seen, bacterial diseases can be treated with antibiotics. The drugs affect the bacterial cells in one way or another (see **Section 2.2.1**). Viruses, however, are not living cells in the conventional sense. Scientists have not yet developed drugs that can affect the virus particles themselves. Instead, antiviral treatments target virus replication. There are a number of different ways in which they can work. They can:

- target the receptors by which viruses recognise their host cells

- target the enzymes that help to translate or replicate the viral DNA or RNA
- inhibit the protease enzymes that enable new virus particles to bud from host membranes.

So far, scientists have not been able to cure viral diseases, but they have reduced the time a person is sick (see **fig A**) and can delay the development of symptoms after infection (e.g. the cocktail of antiretroviral drugs used to treat HIV/AIDS).

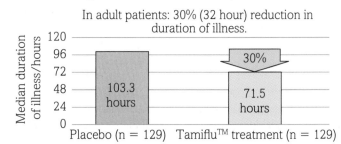

fig A The impact of antiviral medication on the duration of influenza.

Preventing viral disease

Some viral diseases, such as the common cold, are relatively mild and have a very low **mortality rate**. Others, however, are very serious. During 1918–19 an outbreak of influenza killed up to six times more people than the whole First World War. Foot-and-mouth disease has an almost 100% mortality rate in young stock.

In the 2001 UK epidemic of foot-and-mouth disease there was no treatment available and no tests to reliably identify infected animals before they showed symptoms. So all of the cloven-hoofed animals on infected farms were destroyed and burned to try and prevent the spread of the virus to other farms in the area. Over 6 million animals were killed. Veterinary scientists are working on developing sensitive tests to identify infected animals so that control of the disease may be possible in the future without this extensive culling.

fig B Mass culling and burning of possibly infected livestock during the 2001 outbreak of foot-and-mouth disease in the UK was eventually successful in stopping the spread of the disease.

The mortality rate of humans infected with Ebola varies but can be very high with 25–90% of people infected with the disease dying. Mortality depends on the strain of the virus, the health of the infected person and the speed with which they get support and health care. The average mortality rate is around 50%.

Viral diseases like these can be devastating. Because there are no antiviral drugs against most viral infections, disease control focuses on vaccination and reducing the spread of viruses.

Vaccinations

Vaccination plays a major role in the prevention of disease outbreaks. When you are vaccinated against a disease you become immune to it and so will not become infected should you encounter it. Ideally, everyone is vaccinated against serious diseases that may affect them. If an epidemic breaks out and the population is not vaccinated, there is a rush to deliver vaccines to everyone who is not already infected. Usually health care workers, the very young and the elderly are vaccinated first. Unfortunately we have not yet developed fully-tested vaccines against some of the worst viral diseases, such as HIV/AIDS and Ebola. You will learn more about vaccination if you continue to study A level Biology.

Disease control

Understanding the cause of a disease and how it is spread means we can work to control it. Disease control is particularly important when an epidemic occurs. An epidemic is when the levels of people with a particular disease are much higher than expected over a given period of time. When there is a vaccine available, this is the time for mass vaccination of vulnerable people, alongside measures to prevent the spread of disease. In diseases where no vaccine is available, controlling the spread of the disease is key.

Identifying the pathogen early and putting control measures in place can make all the difference to the numbers of people affected (see **fig C**).

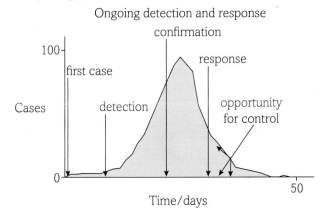

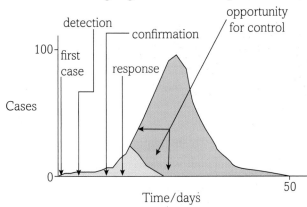

fig C These graphs show the difference that effective detection and response can make to the number of cases that develop.

There are a number of different ways of controlling the spread of a disease. Some are relevant to all diseases, some are only used in more extreme diseases such as Ebola. They include:

* Rapid identification of disease: For example, in West Africa in 2014, it was some time before the Ebola was recognised and effective testing regimes put in place. If the disease is bacterial, it must be identified and an effective antibiotic used.

* Nursing in isolation: This is used for serious infections such as Ebola and *C. difficile* only. It is readily available in countries such as the UK but sealed isolation units are rarely available in developing countries such as those in West Africa. This lack of health infrastructure makes it difficult to isolate people affected by diseases such as Ebola. When ill people are cared for within their families the virus spreads easily. Simple units nursing all infected patients together can help.

* Preventing transmission from one individual to another: Simple measures such as regular hand washing, hand washing before and after every contact with patients by health workers and families, care in handling body fluids and wastes, careful disposal of infected bodily wastes, and frequent disinfecting of surfaces and people are key. Body fluids are very infectious in Ebola cases and good hygiene is vital.

* Sterilising or disposing of equipment and bedding after use: One of the main transmission routes for Ebola at the beginning of the epidemic was through unsterilized needles used in an antenatal clinic.

* The wearing of protective clothing by health workers: When dealing with dangerous and highly infectious viruses such as Ebola, health workers should wear facemasks, gowns, gloves, and goggles to protect the eyes. The slightest contact of infected material with the eyes is enough to lead to infection. The gloves should be washed and disinfected before removal and then the hands washed as well.

* Indentifying contacts: People who have been in contact with infected people need to be monitored so that they can be treated and/or isolated rapidly if they show signs of disease.

Did you know?

Rituals and infection

Many cultures have rituals that are carried out after a death. Family and friends may visit, touch and kiss the body, and they may wash the body and prepare it for burial in the family home. In the case of a disease such as Ebola, the body remains highly infectious after death and so funeral rituals can lead to outbreaks of further infection. It was observed that about a week after the funeral of an Ebola victim, many of the mourners would become sick themselves. Communities accepted that these rituals had to be changed to prevent the spread of Ebola. By sealing bodies in plastic and burying them immediately after death, the spread of the disease was greatly reduced.

In the twenty-first century, in countries such as the UK, we expect to be able to take some medicine and get better if we feel unwell. In an epidemic caused by a virus this isn't always possible due to limited treatment options. If the epidemic is of a potential killer disease, such as flu or Ebola, the pressure to find an effective treatment or vaccine is very high.

The development of new medicines

The development of a new medicine or vaccine takes up to 10 years, involves many different scientists and doctors, and costs millions of pounds. Initial ideas for potential drugs come from a wide range of sources including genome analysis of pathogens, computer modelling, clinical compound banks and medicinal plants. These chemicals have to go through thorough research and testing on cell and tissue cultures, safety analyses and molecular modifications. This is followed by animal testing to ensure the compound works in a whole organism and is safe. This is then followed by three phases of human testing to further ensure safety and that the drug works well. This goes along with complex regulation and licensing procedures until finally, a new drug may reach the doctor's surgery. This process is summarised in **fig D**.

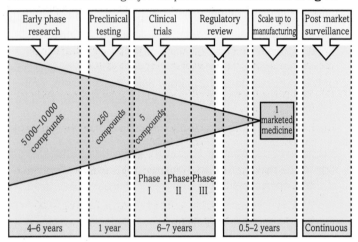

fig D This diagram summarises the main stages in the normal drug development process.

Speeding up the process

When an epidemic develops, some of the final stages of the testing of a new medicine or vaccine may be speeded up to try and save lives and prevent the spread of a deadly disease. Here are two examples.

In 2006 a new strain of H5N1 flu (known as bird flu) caused a global **pandemic.** A vaccine against the new strain was produced very quickly, fast-tracked using existing techniques and technology for producing annual flu vaccines and put through basic clinical trials. They were to be used for health workers if the pandemic hit the UK. The Medical Research Council said they expected the vaccines would give some, if not total, protection whilst a better vaccine was developed. In addition, antiviral medicines such as Tamiflu™ and Relenza™ were stockpiled in spite of concerns that there was incomplete data about their effectiveness. These concerns were raised again when the drugs were used in the 2009 swine flu epidemic and the Cochrane Collaboration, which carries out systematic analysis of the published data on medicines, has since stated that these drugs have not been proven to reduce hospitalisations and serious complication from influenza. You will learn more about influenza in **Book 2**.

The first case of Ebola in this outbreak occurred in late 2013 in West Africa. However, it took until mid-2014 for the world beyond Africa to recognise the size and severity of the outbreak of disease and the speed with which it was spreading. Once the severity of the outbreak was recognised, the World Health Organisation (WHO) and pharmaceutical companies around the world looked for ways to fast-track drugs and vaccines that were already in development and had passed many of the development stages, but which had not completed human trials. In this situation there are two challenges: to make sure the drugs are safe and effective and also to ramp up production to be able to make enough of the medicine or vaccine for it to be useful. Potential treatments included:

- ZMapp™, an experimental drug produced after long-term studies of people who had survived Ebola in previous less-widespread outbreaks. Scientists had genetically modified tobacco plants to produce three antibodies that seem to be associated with surviving the disease. In trials it was effective in treating monkeys, but had not been tried on people. Tiny amounts of the drug were available and used to treat 7 people including African, American, Spanish and British health workers who developed Ebola. Some recovered, but some of the seven died, as you would expect with a disease with around 50% mortality. Scientists are trying to produce more of the substance to run bigger trials on more people.

- Vaccines: Several companies had vaccines in trials that are being fast-tracked for use against Ebola. They are making many doses of the vaccine so that if they are safe to use in humans, many health workers and then people living in epidemic areas can be vaccinated.

Other pharmaceutical companies are supporting the work of companies with drugs and vaccines closest to completion, and are also developing other drugs against the virus itself.

Ethical implications

Historically, doctors and scientists tried out new medicines on themselves, their families or their patients with little or no testing or trials. Today it would be considered completely unethical under normal circumstances to give anyone a medicine or vaccine that had not been through the full process of testing and approval. However, in severe epidemics or pandemics, with thousands of lives at risk, decisions may be made to use drugs that are only part way through the full testing process. Most often this involves drugs that have not completed human trials. Although the media will report these as 'untested', they have in fact already undergone a minimum of five years testing and development, and often will be part way though human trials.

There are a number of factors that have to be evaluated when considering whether a drug should be fast-tracked for use in an epidemic. These include:

- the severity of the disease
- the availability of any other treatments for the disease
- the effectiveness of standard disease control measures in halting the spread of the disease
- transparency about the process and informed consent of those given the treatment
- freedom of choice over participation
- involvement of the affected community – community consent for treatment can be more valuable than individual consent

collection of clear clinical data from the use of new medicines in this situation so an on-going assessment of the safety and efficacy of the drug or vaccine can be made.

Reasons against using untested drugs include:

- Some people simply feel that it is not ethical under any circumstances to use drugs that have not completed full human trials.

- If an untested drug produces unexpected side effects it can make the situation worse

- Deciding who gets the drug or vaccine can be difficult. For example, in a situation such as the Ebola epidemic, local people might feel they were being used as guinea pigs for Western medicine if they are given the medicine, but might feel resentful if only health workers are treated.

- Informed consent is an issue as it depends on a level of education to understand the drug and how it works and also clarity of thought. People who are dying may grasp at straws but their relatives may then blame the treatment for an inevitable death.

- Issues of trust between individuals or communities and health workers, especially if supplies of a new drug are limited.

Did you know?

No epidemic but more ethical decisions

Fungal infections can kill people if their immune system is not working well, for example, in people suffering from diseases such as leukaemia or HIV/AIDS, or those taking immunosuppressant drugs. In the late 1980s the search was on for a new antifungal medicine. Chris Hitchcock and his team at the pharmaceutical company Pfizer set out to design a new molecule that would be more powerful than the existing fungicides and would also kill fungal pathogens resistant to the antifungal drugs available at the time. A molecule known as voriconazole was discovered in 1990 and the long process of development and trialling began.

In 1997 there was a tragedy at the Maccabiah Games, held every four years in Israel. As the Australian team entered the arena over a foot bridge across the very polluted Yarkon River, it collapsed. Over 100 athletes were injured and four died. Three of these deaths were due to a deadly fungal infection picked up from the river. Sasha Elterman, a talented 15-year-old tennis player, was infected with the fungus that day. It attacked her brain and spine and she was given only a 3% chance of surviving. After several months of treatment with every available antifungal medicine, Sasha was still alive – but only just.

Then her medical team read about voriconazole, but it had only just started clinical trials so was a long way from getting a licence. Sasha's doctors got permission to try it as at this stage she had nothing to lose. Without a different treatment she was going to die.

The improvement in Sasha's condition was almost immediate and after 451 days of treatment with the new anti-fungal drug she was fully recovered.

At the opening ceremony of the 2000 Sydney Olympics, Sasha carried the torch into the arena at the head of the Australian Olympic team. She was alive as a result of the use of a new and incompletely trialled medicine. In 2002 voriconazole finally passed all its clinical trials successfully and was licensed. It is still used effectively to treat life-threatening fungal infections today.

In any epidemic situation, the ethical implications of using a fast-tracked and relatively untested drug have to be evaluated at national and international levels. In the US, the Federal Drug Agency decided that an antiviral called peramivir that had not completed testing could be used intravenously in seriously ill patients in the 2009 H1N1 flu epidemic. The WHO recently decided that ZMapp™, which had had no human trials, could be used in the Ebola epidemic in Western Africa and that at least two vaccines could also be fast-tracked through the process for use. The effectiveness of these interventions is yet to be seen, and only then can a full evaluation be made.

Questions

1 Why is the control of the spread of disease particularly important in viral diseases?

2 What might be a disadvantage of giving people a medicine that reduces the symptoms of a viral disease such as flu?

3 (a) Summarise the main ways in which the spread of an infectious disease can be controlled.

   (b) Explain why it was particularly difficult to contain the spread of the Ebola virus in West Africa in 2014.

4 (a) Make a flow chart to show the main stages in the normal drug development process.

   (b) Which stages are most likely to be bypassed if a drug is fast-tracked for use in an epidemic?

5 Suggest why the severity of the disease, the availability of other treatments, and the effectiveness of normal disease control measures are such important factors when evaluating whether a new drug should be tried.

6 Write a paragraph supporting the use of a new drug that has not undergone human trials in the 2014 Ebola outbreak, and a similar paragraph against. In each case evaluate the evidence and put forward scientifically sound opinions.

7 Are the ethical considerations for using an unlicensed treatment the same or different when considering the treatment of a single individual like Sasha Elterman or a community such as those in Sierra Leone during an epidemic? Discuss.

Key definitions

Ebola is a highly infectious viral disease that causes fever and internal bleeding and death in about 50% of cases.

The **mortality rate** is a measurement of the number of deaths in a given population or due to a specific cause.

A **pandemic** is an epidemic that takes place in several countries at once.

EBOLA – A DEADLY VIRUS

In 1976 a new and deadly disease appeared in Africa. Ebola virus causes Ebola, a disease that is so damaging to the body tissues that it has a 25–90% death rate. It is both contagious (spread by contact) and infectious (spread by droplets in the air) and symptoms can appear from 2 to 21 days after the initial infection. Reporting on this viral disease varies greatly…

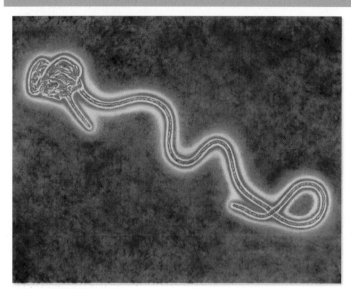

fig A The deadly Ebola virus.

From the *Daily Mail*:

Deadly Ebola virus 'could spread globally' after plane brings it to Nigeria

- Health experts fear other passengers could now be carrying the virus
- It lays dormant in victims for up to three weeks – and 90 per cent die of it

By NICK FAGGE

…The news came as it emerged that an American doctor working for a charity in Liberia had become infected. Dr Kent Brantly, 33, from Texas, had moved to the country for the Samaritan's Purse organisation with his children and wife, Amber, to help contain the disease.

More than 1,000 others have been infected by the virus, which can go unnoticed for three weeks and kills 90 per cent of victims.

Follow us: @MailOnline on Twitter

From the website of the World Health Organization:

Epidemiology and surveillance

WHO continues to monitor the evolution of the Ebola virus disease (EVD) outbreak in Sierra Leone, Liberia and Guinea. The Ebola epidemic trend remains precarious. Between 21 and 23 July 2014, 96 new cases and 7 deaths were reported from Liberia and Sierra Leone. In Guinea, 12 new cases and 5 deaths were reported during the same period. These include suspect, probable and laboratory-confirmed cases. The surge in the number of new EVD cases in Guinea after weeks of low viral activity demonstrates that undetected chains of transmission existed in the community. This phenomenon … calls for stepping up outbreak containment measures, especially effective contact tracing.

From the website of Public Health Wales:

Ebola virus disease: an overview

Ebola virus disease is a serious, usually fatal, disease for which there are no licensed treatments or vaccines. But for people living in countries outside Africa, it continues to be a very low threat.

The current outbreak of the Ebola virus mainly affects three countries in West Africa: Guinea, Liberia and Sierra Leone. …This is the largest known outbreak of Ebola.

So far, there has been just one imported case of Ebola in the UK. Experts studying the virus believe it is highly unlikely the disease will spread within the UK.

What are the symptoms, and what should I do if I think I'm infected?

A person infected with Ebola virus will typically develop a fever, headache, joint and muscle pain, a sore throat, and intense muscle weakness.

These symptoms start suddenly, between two and 21 days after becoming infected, but usually after five to seven days.

If you feel unwell with the above symptoms within 21 days of coming back from Guinea, Liberia or Sierra Leone, you should stay at home and immediately telephone 111 or 999 and explain that you have recently visited West Africa.

These services will provide advice and arrange for you to be seen in a hospital if necessary so the cause of your illness can be determined.

It's really important that medical services are expecting your arrival and calling 111 or 999 will ensure this happens.

Where else will I encounter these themes?

Let us look at the different levels of information given in these pieces of writing, and consider who they are aimed at:

1. The extracts here come from a popular newspaper, the World Health Organization website and the Public Health Wales website.

   a. Which article seems to be the most scientific, and which the least? Explain your answer.

   b. Discuss the different purposes of the three pieces of writing and consider whether you think they are each fit for their purpose.

   c. Each of these extracts shows a bias – they are trying to communicate different things. Comment on what each of the pieces is trying to do in terms of informing the readers.

> Remember that a newspaper has to persuade people to buy copies or pay an online subscription. WHO and Public Health Wales know that anyone visiting their sites has a genuine interest in finding out detailed facts about the topic they are researching!

Now let us examine the biology given in each piece of writing. You already know about viruses and bacteria, so you can answer these questions now. If you are going to continue your biology studies to A level, you may like to revisit these pages after you have learned more about communicable diseases in **Book 2 Topic 6**.

2. Look at the newspaper article and summarise the knowledge about Ebola that you have at the end. How accurate is that information biologically?

3. Summarise what the extract from the Public Health Wales website tells you about Ebola and how you think the virus causes the symptoms of the disease.

4. If you become ill after visiting certain African countries, the Public Health Wales website emphasises the need to inform a doctor or any hospital you attend. Why is this so important?

5. The WHO extract gives little or no information about the Ebola virus itself. What is the focus of this article? Why is this information also important biologically?

Activity

Research is key in preventing the spread of viruses like Ebola.

Find out as much as you can about the Ebola virus, focusing on how it is spread and the way it infects and takes over the cells of the body.

Think carefully about which stage of the viral life cycle you would target to try and prevent the spread of the disease.

Imagine you have to bid for funding to carry out your research. Put together a poster presentation summarising the problem you have identified with Ebola, explaining what you want to research and why you should receive the funding.

> Be clear about the different types of viruses and the ways they reproduce in cells.
>
> Refer to the information in the Public Health Wales extract and visit the original website as well as others and use the information in your textbook. Think very carefully about where and when a virus might be vulnerable to attack during its reproductive cycle.

1 The table below refers to some features of prokaryotic and eukaryotic cells. If the feature is present, place a tick (✔) in the appropriate box and if the feature is absent, place a cross (✗) in the appropriate box.

Feature	Prokaryotic cell	Eukaryotic cell
Nuclear envelope		
Cell surface (plasma) membrane		
Mitochondria		
Golgi apparatus		

[Total: 4]

2 An analysis of the large organic molecules found in a prokaryotic cell was made. The dry mass of the cell is the mass of the cell not including water. The results of the analysis are shown in the table below.

Molecule	Percentage of total dry mass of the cell/%	Number of molecules per cell	Number of different types of molecule
Protein	55.0	2 360 000	1050
Lipid	9.1	22 000 000	4
Glycogen	2.5	4360	1
DNA	3.1	2	1
RNA	20.5	262 480	463

(a) The molecular mass of a substance is the mass of one molecule of that substance. Using information in the table, state which of the molecules has the largest molecular mass.
Give an explanation for your answer. [2]

(b) Glycogen and protein molecules are both polymers. Explain why there is only one type of glycogen molecule but there are many types of protein molecule. [2]

(c) Explain why many different RNA molecules are found in a cell. [2]

[Total: 6]

3 The table below refers to some of the stages involved in Gram staining and the appearance of Gram-negative and Gram-positive bacteria after each stage. Complete the table by writing the most appropriate word or words in the empty boxes.

Stage of Gram staining	Appearance of Gram-negative bacteria	Appearance of Gram-positive bacteria
Cells heat fixed onto slide	Colourless	Colourless
Slide flooded with crystal violet		
Slide flooded with Gram's iodine		
Slide rinsed with alcohol or acetone		
Slide counterstained with safranin/carbol fuchsin		

[Total: 4]

4 The table below refers to features of λ (lambda) phage, tobacco mosaic virus (TMV) and human immunodeficiency virus (HIV). Complete the table by writing the most appropriate word or words in the boxes.

Feature	λ-phage	TMV	HIV
Type of nucleic acid			
Shape of protein coat			

[Total: 6]

5 The table below refers to some structures of microorganisms. Complete the table by writing the name of the type of microorganism possessing each structure in the empty boxes.

Structure	Type of microorganism
Nucleus	
Capsid	
Flagellum	
Peptidoglycan (murein) cell wall	

[Total: 4]

6 Prokaryotes, mitochondria and chloroplasts have many features in common.

(a) (i) The diagram below shows a mitochondrion. Two of the features labelled are typical of prokaryotes. Place a tick (✓) in each of the **two** boxes that correctly identify these features.

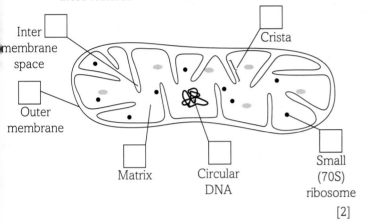

Inter membrane space

Crista

Outer membrane

Matrix

Circular DNA

Small (70S) ribosome

[2]

(ii) The table below shows some features of mitochondria. If the feature is also present in chloroplasts, place a tick (✓) in the box to the right of that feature and if it is absent, place a cross (✗) in the box.

Features present in mitochondria	Feature present (✓) or absent (✗) in chloroplasts
Surrounded by a double membrane	
Crista present	
Circular DNA	
Matrix	
Glycogen granule	
Stalked particles	

[3]

(b) Bacteria can be identified and classified by looking for certain features. Using the information in the passage below, label the five bacteria with the correct letter.

Bacterium P has a single flagellum to enable it to move whilst bacterium Q has several flagella.

Only bacterium R has visible plasmids and bacterium S has an infolding of its cell surface membrane.

Bacterium T has a slime capsule.

Bacterium ....   Bacterium ....

Bacterium ....   Bacterium ....

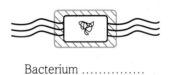

Bacterium ....

[4]

[Total: 9]

CHAPTER

2.3 Eukaryotic cell division – mitosis

Introduction

Normal growth and division in cells occurs in a cycle that is controlled by literally hundreds of genes acting in different ways to stimulate or repress the process. Cells which mutate are usually removed by apoptosis, programmed cell death brought about at least in part by the action of the lysosomes. But sometimes cells go wrong – as they divide they develop mutations that eventually mean they do not respond to the normal signals controlling the cell cycle. This potentially results in one of the many cancers that can affect almost every tissue and organ in the body. Identifying these changes and controlling them, or preventing them from happening in the first place, is the focus of much research.

In this chapter you will be looking at mitotic cell division in eukaryotic cells. The cell cycle is of great importance and you will discover the main stages of the cycle, and how the speed of the cycle varies at different ages and in different tissues.

You will learn the stages by which cell division – mitosis – takes place. By looking at the way the chromosomes replicate and divide in a graceful 'dance', followed by the rest of the cytoplasm of the cell, you will come to understand how mitosis results in two identical daughter cells. You will carry out practical investigations using plant meristems to see the stages of mitosis under the microscope.

You will also consider the importance of mitosis in living organisms – it is the basis of asexual reproduction in many animals, plants and fungi. Mitosis produces offspring that are identical to the single parent, and it can result in enormous numbers of offspring being produced at one time.

Mitosis is also important for the repair of damaged tissues and for normal growth from infancy to adulthood, and you will look at how this growth can be observed and measured. Growth in animals stops when they reach maturity. In plants, growth continues throughout life, a pattern revealed in the trunks of trees when they are felled.

All the maths you need

- Carry out calculations using numbers in standard and ordinary form (*e.g. use of magnification*)
- Use scales for measuring (*e.g. measuring sizes of cells at different stages of growth*)
- Find arithmetic means (*e.g. measuring sizes of cells at different stages of the cell cycle*)
- Make order of magnitude calculations (*e.g. use and manipulate the magnification formula:*
  $$magnification = \frac{size\ of\ image}{size\ of\ real\ object}$$
)
- Use and manipulate equations, including changing the subject of an equation (*e.g. magnification*)

What have I studied before?

- Asexual reproduction in living things
- The basic process of mitosis, including the idea of the cell cycle
- That cancer is the result of changes in cells that lead to uncontrolled growth and division
- The ultrastructure of eukaryotic cells including the nucleus, nuclear membrane, chromosomes, centriole, etc.
- The way in which DNA replicates in the nucleus
- Gene mutations

What will I study later?

- Meiosis – the process of cell division that reduces the number of chromosomes in the nucleus to form the gametes
- The early development of a mammalian embryo to the blastocyst stage, that involves rapid mitosis
- Fertilisation in plants
- The growth of a bacterial colony that will involve understanding binary fission (A level)
- Measuring growth of bacterial colonies (A level)
- The immune response of the body including clonal selection and the rapid mitosis that results in the production of plasma cells and T killer cells as well as T and B memory cells (A level)
- Stem cells in animals and plants (A level)
- The control of cell differentiation (A level)
- Plant responses to environmental stimuli that depend on cell division and growth

What will I study in this chapter?

- The cell cycle as a regulated process made up of interphase, mitosis and cytokinesis, in which cells divide to produce two identical daughter cells
- The replication and division of the genetic material in the main stages of mitosis
- The importance of mitosis in growth, repair of damaged or ageing tissues, and asexual reproduction to produce offspring that are identical to the one parent
- Techniques to make a temporary squash preparation of the cells of a root tip where active mitosis is taking place

By the end of this section, you should be able to...

● describe the cell cycle as a regulated process in which cells divide into two identical daughter cells

● describe the three main stages of the cell cycle as interphase, mitosis and cytokinesis

● explain what happens to the genetic material during the cell cycle

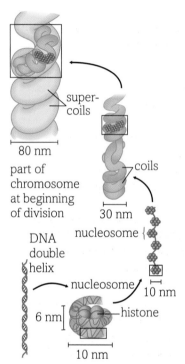

fig A Histones play an important role in the organisation of DNA into orderly chromosomes that can be replicated.

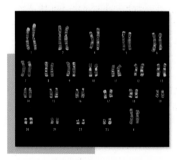

fig B This female human karyotype shows the 22 pairs of autosomes and one pair of sex chromosomes found in every healthy human cell, except the eggs and sperm.

One of the most awe-inspiring processes of life is the way in which organisms reproduce. Like begets like – buttercups produce new buttercups, *Amoeba* produce more *Amoeba* and liver cells generate more liver cells. Most new biological material comes about as a result of the process of nuclear division known as **mitosis**, followed by the rest of the cell dividing. **Asexual reproduction** – the production of genetically identical offspring from a single parent cell or organism – and growth are both the result of mitotic cell division. The production of offspring by **sexual reproduction** is also largely dependent on mitosis to produce new cells after the gametes (sex cells) have fused. In mitosis the chromosomes of a cell are duplicated and the genetic information is then equally shared out between the two daughter cells that result. The formation of the sex cells involves a different process of nuclear division called **meiosis** (see **Section 2.4.1**).

What are chromosomes?

Eukaryotic cell division involves replicating the chromosomes that carry the genetic information. A chromosome is made up of a mass of coiled threads of DNA and proteins. If a chromosome were as long as five consecutive letters on this page, the DNA molecule it contained would stretch the length of a football pitch or more. In a cell that is not actively dividing, the chromosomes are translucent to both light and electrons so we cannot see them easily or identify them as individual structures. When the cell starts to actively divide, the chromosomes condense – they become much shorter and denser. They then take up stains very readily. This is the basis of the name 'chromosome' or 'coloured body' and as a result at this stage of the process we can identify individual chromosomes.

When the DNA molecules condense, they have to be packaged very efficiently. This is achieved with the help of positively charged basic proteins called **histones**. The DNA winds around the histones to form dense clusters known as **nucleosomes** (see **fig A**). These then interact to produce more coiling and then supercoiling to form the dense chromosome structures you can see through the microscope in the nucleus of a dividing cell. In the supercoiled areas the genes are not available to be copied to make proteins.

The cells of every different species possess a characteristic number of chromosomes – in humans this is 46. These chromosomes occur in matching pairs, one of each pair originating from each parent. In mitosis the two cells that result from the division must both receive a full set of chromosomes. So before a cell divides it must duplicate the original set of chromosomes. During mitosis these chromosomes are divided equally between the two new cells so that each has a complete and identical set of genetic information. During the active phases of cell division the chromosomes become very coiled and condensed. In this state they can be photographed to produce a special display or **karyotype** (showing all the chromosomes of the cell).

The cell cycle

Cells divide on a regular basis to bring about growth and asexual reproduction. They divide in a sequence of events known as the **cell cycle**, which involves several different phases (see **fig C**). **Interphase** is a period of non-division when the cells increase in mass and size, carry out normal cellular activities and replicate their DNA ready for division. This is followed by mitosis, a period of active division, and cytokinesis when the new cells separate. The length of the cell cycle is variable. It can be very rapid, taking 24 hours or less, or it can take a few years.

Phases of the cell cycle

- G_1 (gap 1) is the time between the end of the previous round of mitotic cell division and the start of chromosome duplication. The cell assimilates material, grows and develops. This is the time that is most variable. In actively dividing cells, G_1 is very short – a matter of hours or days, but in other cells, it can be months or even years.

- S is the stage when the chromosomes replicate and become double stranded **chromatids** ready for the next cell division.

- G_2 (gap 2) is the time that the organelles and other materials needed for cell division are synthesised – before a cell can divide, it needs two of everything.

- M is mitosis when the cells are actively dividing.

- C is cytokinesis, the final stage of the cell division when the new cells separate.

In multicellular organisms the cell cycle is repeated very frequently in almost all cells during development. However, once the organism is mature, it may slow down or stop completely in some tissues. The cell cycle is controlled by a number of chemical signals made in response to different genes. This control is brought about at a number of checkpoints where the cell cycle moves from one phase to the next. The control chemicals are small proteins called **cyclins**. These build up and attach to enzymes called **cyclin-dependent kinases (CDKs)**. The cyclin/CDK complex that is formed phosphorylates other proteins, changing their shape and bringing about the next stage in the cell cycle. Examples include the phosphorylation of the chromatin in the nucleus, which results in the chromosomes becoming denser, and the phosphorylation of some of the proteins in the nuclear membrane, which leads to the breakdown of the nuclear membrane structure during cell division.

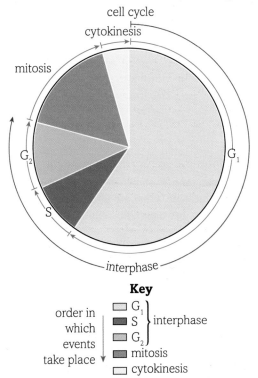

fig C The phases of the cell cycle. In very actively dividing tissue the cycle is repeated as fast as possible, whilst in other tissues the time between successive divisions may be years.

Did you know?

Permanent cells

There are some cells that do not enter the cell cycle once they have formed – they have to last a lifetime. They are known as permanent cells. Examples include nerve cells, the light sensitive cells of the retina, the transparent cells of the lens of the eye and the cardiac muscle – the muscle that makes up your heart.

Questions

1 Why do chromosomes only become visible as a cell goes into mitosis?

2 If a culture of cells is dividing every 48 hours, how long would you expect the different stages of the cycle to take?

Key definitions

Mitosis is the process by which a cell divides to produce two genetically identical daughter cells.

Asexual reproduction is the production of genetically identical offspring from a single parent or organism.

Sexual reproduction is the production of offspring that are genetically different from the parent organism or organisms by the fusing of two sex cells (gametes).

Meiosis is a form of cell division in which the chromosome number of the original cell is halved, leading to the formation of the gametes.

Histones are positively charged proteins involved in the coiling of DNA to form dense chromosomes in cell division.

Nucleosomes are dense clusters of DNA wound around histones.

A **karyotype** is a way of displaying an image of the chromosomes of a cell to show the pairs of autosomes and sex chromosomes.

The **cell cycle** is a regulated process of three stages (interphase, mitosis and cytokinesis) in which cells divide into two genetically identical daughter cells.

Interphase is the period between active cell divisions when cells increase their size and mass, replicate their DNA and carry out normal metabolic activities.

A **chromatid** is one strand of the replicated chromosome pair that is joined to the other chromatid at the centromere.

Cyclins are small proteins that build up during interphase and are involved in the control of the cell cycle by their attachment to cyclin-dependent kinases.

Cyclin-dependent kinases (CDKs) are enzymes involved in the control of the cell cycle by phosphorylating other proteins, activated by attachment to cyclins.

A cell is in the interphase stage of the cell cycle for much of its life. This used to be called the resting phase, but nothing could be further from the truth. During interphase the normal metabolic processes of the cell continue and new DNA is produced as the chromosomes replicate. New proteins, cytoplasm and cell organelles are synthesised so that the cell is prepared for the production of two new cells. ATP production is also stepped up at times to provide the extra energy needed as the cells divide. Once all that is needed is present and the parent cell is large enough, interphase ends and mitosis begins.

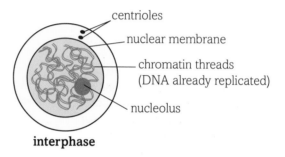

interphase

The stages of mitosis

During the process of cell division the chromosomes duplicated during interphase are divided up with the remaining contents of the cell so two identical daughter cells are formed. Walther Flemming (1843–1905), a German cytologist, was the first to describe what is sometimes called the 'dance of the chromosomes'. It refers to the complex series of movements that occur during cell division as the chromosomes jostle for space in the middle of the nucleus and then pull apart to opposite ends of the cell. The events of mitosis are continuous, but as in the case of so many biological processes it is easier to describe what is happening by breaking events down into phases. These are known as **prophase**, **metaphase**, **anaphase** and **telophase**.

Prophase

Before mitosis begins the genetic material has been replicated to produce exact copies of the original chromosomes. By the beginning of prophase both the originals and the copies are referred to as chromatids. In prophase the chromosomes coil up, can take up stains and become visible. Each chromosome at this point consists of two daughter chromatids that are attached to each other in a region known as the **centromere**. The nucleolus breaks down and the centrioles begin to pull apart to form the spindle (see **Section 2.1.3**).

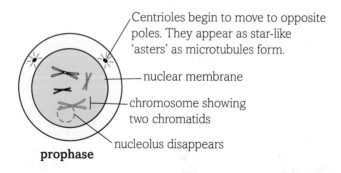

prophase

Metaphase

The nuclear membrane has broken down and the centrioles have moved to opposite poles of the cell, forming a set of microtubules between them that is known as the spindle. The chromatids appear to jostle about for position on the **metaphase plate** or **equator** of the spindle during metaphase. They eventually line up along this plate, with each centromere associated with a microtubule of the spindle.

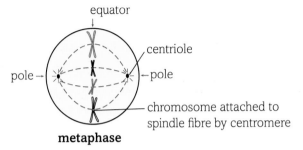

metaphase

Anaphase

The centromeres that have linked the two identical chromatids split, and from then on the chromatids act as completely separate entities. They effectively become new chromosomes. The chromatids from each pair are drawn, centromere first, towards opposite poles of the cell. This separation occurs quickly, taking only a matter of minutes. At the end of anaphase the two sets of chromatids have been separated to opposite ends of the cell. The chromatids cannot move on their own. They rely on the microtubules of the spindle to allow them to move. The spindle was for many years envisaged as a structure running from one end of the cell to the other. It is now known to be made up of overlapping microtubules containing contractile fibres, which are similar to those in animal muscles. Contraction of the overlapping fibres causes the movement of the chromatids. This is an energy-using process, and the energy is supplied by cell respiration.

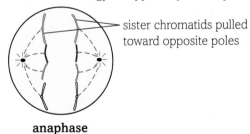

anaphase

Telophase

During telophase the spindle fibres break down and nuclear envelopes form around the two sets of chromosomes. The nucleoli and centrioles are also re-formed. The chromosomes begin to unravel and become less dense and harder to see.

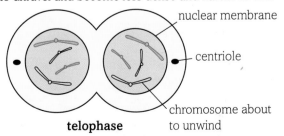

telophase

Cytokinesis

The final phase of the cell cycle is the division of the cytoplasm, sometimes referred to as **cytokinesis**. In animal cells a ring of contractile fibres tightens around the centre of the cell rather like a belt tightening around a sack of flour. These fibres seem to be the same as those found in animal muscles. They continue to contract until the two cells have been separated. In plant cells the division of the cell occurs rather differently, with a cellulose cell wall building up from the inside of the cell outwards. In both cases the end result is the same – two identical daughter cells are formed, which then enter interphase and begin to prepare for the next cycle of division.

(a)

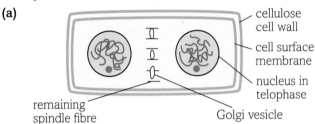

Some spindle fibres remain and guide Golgi vesicles to the equator of the cell.

The vesicles enlarge and fuse together, forming a cell plate.

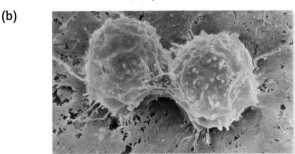

The basic structure of the cell walls forms within each vesicle, and the vesicles fuse to join the cell wall together. Small gaps left between the vesicles form plasmodesmata (see **Section 2.1.5**).

(b)

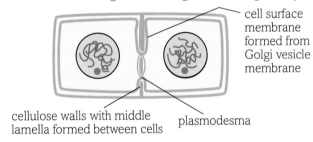

fig A The final stages of the cell cycle in (a) a plant cell; and (b) an animal cell.

Mitosis is the source of all the new cells needed for organisms to grow and to replace worn out cells. It is also the method by which organisms undergo asexual reproduction.

Did you know?

Observing mitosis

The discovery of mitosis depended on the development of the microscope. Walther Flemming published his work on mitosis in 1882. Flemming had also discovered the presence of chromosomes in the cell by using dyes that were taken up by the genetic material. A Belgian scientist, Edouard van Beneden (1846–1910), discovered chromosomes at much the same time. Flemming had not come across Mendel's work on inheritance and so he did not make the connection between what he was seeing and genetic inheritance. In spite of this, Flemming's discoveries are widely regarded as some of the most important work in cell biology.

You can observe mitosis relatively easily in the cells of rapidly dividing tissues such as the meristem at a growing root tip. Using a dye such as acetic orcein, which stains the chromosomes, you can make a temporary tissue squash preparation showing the stages of mitosis. You can also observe mitosis in living tissue, and dramatic recordings of the activity of the chromosomes have been made using time-lapse photography. This has moved our understanding forward considerably. Viewing of the movements of the cell contents during mitosis shows it as a dynamic process in a way which cannot be achieved on the printed page and explains why it is called the 'dance of the chromosomes'.

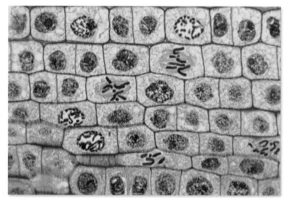

fig B Stained section of a root tip squash showing cells in different stages of the cell cycle, including active mitosis.

Questions

1   Summarise the stages of mitotic cell division in animal cells.

2   Explain why root tips are particularly suitable material to use for preparing slides to show mitosis.

Key definitions

Prophase is the first stage of active cell division where the chromosomes are coiled up and consist of two daughter chromatids joined by the centromere. The nucleolus breaks down.

Metaphase is the second stage of active cell division where a spindle of overlapping protein microtubules forms and the chromatids line up on the metaphase plate.

Anaphase is the third stage of active cell division where the centromeres split so chromatids become new chromosomes. They are moved to the opposite poles of the cell, centromere first, by contractions of the microtubules of the spindle.

Telophase is the fourth stage of active cell division where a nuclear membrane forms around the two sets of chromosomes, the chromosomes unravel and the spindle breaks down.

The **centromere** is the region where a pair of chromatids are joined and which attaches to a single strand of the spindle structure at metaphase.

The **metaphase plate (equator)** is the region of the spindle in the middle of the cell along which the chromatids line up.

Cytokinesis is the final stage of the cell cycle before it enters interphase again – division of the cytoplasm at the end of mitosis to form two independent, genetically identical cells.

Asexual reproduction

By the end of this section, you should be able to...

● explain the importance of mitosis in asexual reproduction

As you have already seen, mitosis is the basis of asexual reproduction. Asexual reproduction involves only one parent individual and it results in genetically identical individuals or **clones**. It has many advantages for an organism. It does not rely on finding a mate and can give rise to large numbers of offspring very rapidly. It also has one big disadvantage – the offspring are almost all genetically identical to the parent organism. This becomes a problem when living conditions change in some way. The introduction of a new disease to an environment, a change in the temperature or human intervention can cause the total destruction of a group of genetically identical organisms, because if one cannot cope, neither can all the others. Many species of plants and fungi undergo both sexual and asexual reproduction as a matter of course. For example, plants may reproduce sexually by flowering, but they also reproduce asexually by methods including bulbs, corms, tubers, rhizomes, runners and suckers. Single-celled animals such as *Amoeba* divide by mitosis to produce new individuals.

Strategies for asexual reproduction

There are a variety of strategies for asexual reproduction, all of which are dependent on mitosis and some of which are outlined below.

Producing spores

Sporulation involves mitosis and the production of asexual spores that are capable of growing into new individuals. These spores can usually survive adverse conditions, and are also easily spread over great distances. This form of asexual reproduction is most common in fungi and plants such as mosses and ferns.

Regeneration

Regeneration constitutes a very dramatic form of asexual reproduction, occurring when organisms replace parts of the body that have been lost. For example, many lizards shed their tails when attacked and then grow another. This is known as regeneration. Some organisms manage an even more spectacular form of regeneration – they can reproduce themselves asexually from fragments of their original body, a process known as **fragmentation**. For example, certain starfish attack and eat oysters. To protect oyster beds from destruction oyster fishermen have attempted to destroy the starfish, often by chopping them up and throwing them back into the sea. This failed dismally as each fragment can regenerate to form another starfish hungry for oysters.

This type of cloning occurs naturally – some members of groups as diverse as fungi, flatworms, filamentous algae and sponges fragment and then regenerate as a regular method of reproducing. An adaptation of this ability has been developed to allow artificial cloning of plants.

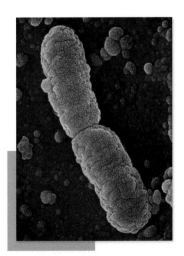

fig A Fission is an effective reproductive strategy for prokaryotes such as this bacterium.

Did you know?

Bacteria reproduce by a process known as binary fission. Bacteria are capable of enormous increases in numbers under ideal conditions, when they may divide every 20 minutes - one of the shortest known cell cycles. Although fission is limited as a reproductive strategy in the world of multicellular organisms, a similar method is used in cell reproduction for growth and repair in all living things.

fig B This lizard lost its tail escaping a predator. The tail has not just healed, it is actually regenerating using mitosis.

Producing buds

Budding in a reproductive sense does not mean the production of buds containing flowers or leaves. In reproductive budding there is an outgrowth from the parent organism that produces a smaller but identical individual, produced purely by mitotic cell division. This 'bud' eventually becomes detached from the parent and has an independent existence. Yeast cells, which are single-celled fungi, reproduce by budding. Budding is relatively rare in the animal kingdom. A good example of an animal budding is in *Hydra* (see **fig C**). Asexual budding is only part of the reproductive strategy of *Hydra* – they reproduce sexually as well.

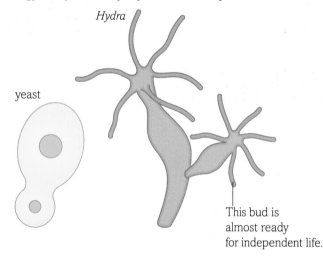

fig C Budding in yeast cells and *Hydra*. Even in single-celled yeast the new organism is much smaller than the parent.

New plant structures

Vegetative propagation is in some ways a more sophisticated version of reproductive budding and occurs in flowering plants. A plant forms a structure that develops into a fully differentiated new plant, which is identical to the parent, and eventually becomes independent. The new plant may be propagated from the stem, leaf, bud or root of the parent, depending on the type of plant. It involves only mitotic cell division. Vegetative propagation often involves perennating organs. These contain stored food from photosynthesis and can remain dormant in the soil to survive adverse conditions. They are often not only a means of asexual reproduction, but also a way of surviving from one growing season to the next. Examples include bulbs, corms, runners, suckers, rhizomes, stem tubers and root tubers.

Vegetative propagation is easily exploited by human gardeners to produce new plants. Splitting daffodil bulbs, removing new strawberry plants from their runners and cutting up rhizomes are all easy ways of increasing plant numbers cheaply. As an added advantage the new plants are all clones that will have exactly the same characteristics as their parents, so they will be the same colour or have the potential to produce fruit that is just as good.

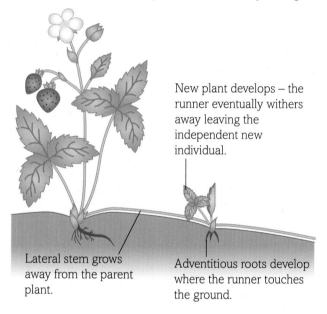

New plant develops – the runner eventually withers away leaving the independent new individual.

Lateral stem grows away from the parent plant.

Adventitious roots develop where the runner touches the ground.

fig D Asexual reproduction in strawberry plants results in identical clones.

Gardeners and farmers take asexual reproduction in plants one step further when they take cuttings. They induce fragmentation artificially. This involves taking a small piece of a plant – often part of the shoot – and planting it to grow on and develop by mitosis into another entire identical plant.

Did you know?

The dangers of cloning crops

Cloning crop plants allows farmers to produce large numbers of productive crops very quickly. Clones are commonplace - potato crops, bananas and grape vines are examples of cloned crops - but because clones are genetically identical, if a disease affects one plant in a crop it will affect them all. The potatoes that were destroyed by fungal blight in the Irish potato famine of the 1840s were all clones. They were lumper potatoes, which had no blight resistance. There are more varieties of modern potatoes, but most potato plants are still clones. Bananas are the clearest example of a global monoculture resulting from artificial asexual reproduction. In the early twentieth century, almost all banana plantations grew a clone or cultivar called Gros Michel. In the 1950s, a fungal disease called Panama disease wiped out almost all the banana plantations in South America and Africa. Now most banana plantations use the Cavendish variety, another banana clone, which so far is proving disease resistant. But it only takes a new disease such as black sigotoka to leave us all without the bananas we love to eat.

Asexual reproduction is common even in complex plants partly because they maintain areas of unspecialised dividing cells throughout their life. In more complex animals, where the cells tend to become specialised, asexual reproduction is much less common.

Did you know?

Komodo dragons and virgin births

Most vertebrates do not reproduce asexually. When they do it is known as **parthenogenesis**. This has been seen in about 80 species of vertebrates, including snakes, fish, a monitor lizard and a turkey. In spite of this, parthenogenesis has always been thought to be very rare in vertebrates. Yet new evidence suggests that parthenogenesis is not as rare as was thought. Komodo dragons are the largest land lizards on Earth. Their natural habitat is Indonesia where there are fewer than 4000 animals remaining. So the Komodo dragons in captivity are part of an important breeding programme. However, two different zoos in the UK, Chester Zoo and London Zoo, have reported that live, apparently healthy young dragons have hatched from eggs laid by females in the complete absence of any male dragons. The females were not related in any way, so it was not a rare family mutation. The young dragons have been DNA tested and their genetic make-up shows they come only from their mother. Both females have also bred sexually. Female Komodo dragons have one W and one Z chromosome, while males have two Zs. Each egg carries either a W or a Z. When parthenogenesis takes place, the single chromosome is duplicated. Any eggs with WW will not develop so there are no parthenogenic female babies, but ZZ eggs can develop into normal male baby lizards.

Richard Gibson is in charge of reptiles at London Zoo and an international expert in Komodo dragons. Richard feels that the arrival of these parthenogenic dragons should lead scientists to rethink their ideas on how common the process of parthenogenesis is, at least in reptiles. The ability to reproduce asexually as well as sexually could have evolved so that animals stranded in an isolated situation can nevertheless breed. If so it would be a very useful adaptation indeed – and therefore it would not be surprising if it was relatively common.

Until recently, all the known vertebrate parthenogenic births had been to animals in captivity. However, in 2012, Warren Booth and his team at the University of Tulsa in the USA used DNA genotyping to demonstrate the first documented cases of parthenogenesis in two closely related species of wild pit viper snakes. This throws open the whole question of the role of parthenogenesis in the lives of reptiles and many other species. There is a lot more research to be done.

fig E This small male snake is the result of parthenogenesis – he was born to the wild mother snake with no male input into his genetic make-up.

Questions

1. Based on the information in this section and further research, make a table to summarise the forms of asexual reproduction most commonly seen in eukaryotic organisms.

2. More living organisms result from asexual reproduction than sexual reproduction. Do you think this statement is accurate? Explain your response.

3. New observations can change long-held scientific ideas. Why have the Komodo dragon hatchlings and the wild-born parthenogenic pit vipers forced scientists to rethink their ideas about asexual reproduction in vertebrates?

Key definitions

Clones are genetically identical offspring produced as a result of natural or artificial asexual reproduction.

Binary fission is the splitting of one individual to form two new individuals.

Sporulation is the process involving mitosis in the production of asexual spores that can grow into new individuals.

Regeneration is the use of mitosis to regrow a body part that has been lost.

Fragmentation is the use of mitosis to regenerate a whole organism from a fragment of the original.

Budding is the production by mitosis of an outgrowth from the parent organism that develops into a small independent organism.

Vegetative propagation is the process by which a plant forms a structure by mitosis that develops into a fully differentiated, genetically identical new plant.

Parthenogenesis is the process by which an unfertilised egg cell develops into a new individual.

By the end of this section, you should be able to...

● explain the importance of mitosis in growth and repair

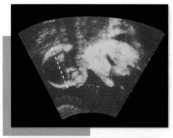

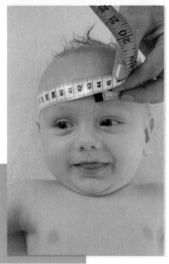

fig A Head circumference is measured at intervals as a child grows, starting before birth, to check that all is well. However, an increase in circumference does not always indicate growth.

Mitosis is not just about asexual reproduction. It plays a vital role in growth as well. Everyone is familiar with the concept of growth, but defining it in a biological sense is not so easy. Growth is a permanent increase in the number of cells, or in the mass or size of an organism. There are three distinct aspects of growth. They are cell division, assimilation and cell expansion. Cell division, or mitosis, is the basis of growth. Once cells have divided they usually get larger before dividing again. The resources needed to produce new cell material come from photosynthesis in plant cells, from feeding in animal cells and from nutrient absorption in fungi. This is what is meant by assimilation, and when these materials are incorporated into cells the result is cell expansion. Cells can expand in other ways, for example by taking in water, but this increase in size may be only temporary. So growth is defined as involving a *permanent* increase in cell number, size or mass – or all three.

How is growth measured?

The measurement of growth is important both scientifically and medically. Growth may be affected by factors such as the availability of food, temperature and light intensity as well as the genetic make-up of the organism. Unfortunately the measurement of growth is not at all easy. Linear dimensions, such as height or head circumference (see **fig A**), can be very deceptive – cake mixture will increase in both height and circumference as it cooks, but it has not grown. Measuring mass also has its problems – the water content of the cells may vary greatly, particularly in fungi and plants, and more complex animals will have varying quantities of faecal material and urine held in their bodies.

Because growth involves an increase in the cell content of an organism, mass is the best and most commonly used measure of growth. However, the water content of organisms can vary greatly so **dry mass** is the most accurate way of measuring growth. The dry mass is the mass of the body of an organism with all the water removed from it. This gives an accurate picture of the amount of biological material present, but has one major drawback. If you remove all the water from an organism you kill it, so that further growth cannot be measured. To get useful results from dry mass measurements you need to grow large samples of genetically identical organisms under similar conditions, then take random samples and dry them to a constant dry mass. This method is very useful for plants, fungi and bacteria, but has obvious limitations for animals. It is not easy or ethical to maintain large colonies of genetically identical vertebrates and then kill and dry them, for example. This means that in most cases scientists use less reliable indicators such as height and wet mass to measure growth when working with animals.

Growth patterns

In spite of the difficulties in measuring growth, we have a good picture of the patterns of growth of many organisms. Growth curves show growth throughout the life of an organism, including when most growth takes place. The growth curve is very similar for most organisms (see **fig B**). In many animals, after an initial relatively slow start there is a rapid period of growth until maturity is reached, when growth slows down and may stop. In most land animals, growth stops completely with maturity because size is limited by the weight of the animal and the ability of its muscles to move it against gravity. In plants, growth often continues throughout life, and the same is true for marine animals, where the mass of the body is supported by the water. This pattern – even when it stops at maturity – is known as continuous growth.

Not all organisms undergo continuous growth. Insects grow in a series of moults. They shed one exoskeleton and then, while the new exoskeleton is soft, they expand the body by taking in air or water and 'grow'. Once the new skeleton has hardened, the air or water can be released and there is room for the tissues of the insect to increase in size and mass. This is known as discontinuous growth. If length is measured the insect appears to grow in a series of steps.

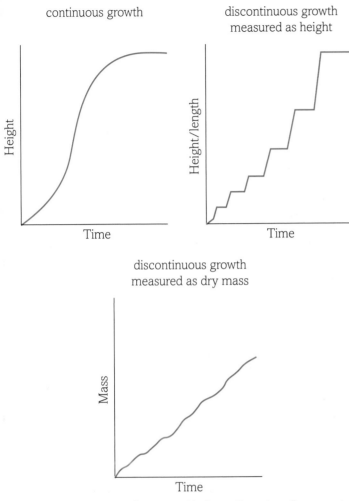

fig B Growth curves are usually measured by linear dimensions. If wet mass is used, the discontinuous growth curve becomes even more pronounced.

The development of the embryo is the time when the largest amount of growth, measured as a percentage of body mass, occurs. Different parts of the organism can grow at very different rates. For example, in the human embryo the nervous system and the head grow much faster than some other areas. Later in life – through puberty, for example – the head stops growing while the long bones and the rest of the body continue. Right at the beginning of life, mitosis takes place at a very rapid rate.

Growth in plants

In plants mitosis continues throughout life. It takes place in regions known as the **meristems**. These are areas just behind the tip of the stem or root where the cells continue to divide actively throughout the life of the plant. After the cells have divided they absorb water into their vacuoles and elongate rapidly before the cellulose cell wall becomes more rigid. These areas of plant growth are particularly sensitive to a variety of stimuli such as light and gravity.

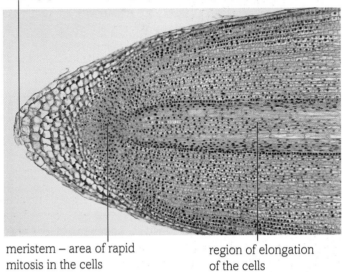

fig C The apical meristems in the root tips and shoots are the sites of growth in a plant.

Mitosis and repair

However, even in organisms where growth slows down or stops completely at maturity, mitosis does not stop. Cells are continually becoming worn out and being replaced by mitotic divisions. In some tissues mitosis occurs at a rapid rate all the time. For example, the entire surface of your skin is replaced every 28 days so mitosis is taking place in skin cells all the time. Each red blood cell only lasts 120 days in the blood so mitosis also occurs rapidly in the red bone marrow of the flat bones such as your ribs to keep up with the demand. This continues until the onset of senescence or old age, when mitosis occurs less frequently and the cells dying begin to outnumber the new cells being formed. When this process reaches a certain point, death of the whole organism will occur.

If the skin is damaged, rapid mitosis is triggered in the cells around the wound. They regenerate the lost skin tissue and heal the wound. Plants can also produce scar tissue to seal damage to their bark. Mitosis is vital for the repair of worn out or damaged tissues.

Questions

1 How does the role of mitosis differ in the processes of growth and repair?

2 Why are the curves for discontinuous growth so different when measured as height and as dry mass?

Key definitions

Dry mass is the mass of the body of an organism with all the water removed from it.

The **meristem** is the region of mitosis and growth in a plant shoot or root.

CANCER – MITOSIS OUT OF CONTROL

In 2013 an estimated 8.2 million people around the world died from different forms of cancer. The article below suggests answers to some common questions about cancer.

What is cancer?

Cancer is an umbrella term that covers more than 200 different diseases. However, most cancers have some features in common:

- Cancer cells do not respond to the normal mechanisms that control cell growth – they divide rapidly to form a mass of abnormally growing cells called a tumour.
- Malignant tumours split and release small groups of cells into the blood and tissue fluid. They circulate and often lodge in different areas of the body, forming secondary tumours.
- Cancer cells divide more rapidly and usually live longer than normal cells.

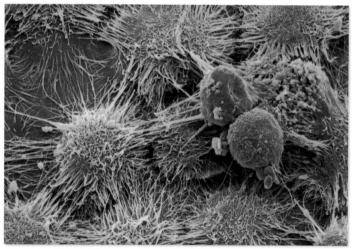

fig A Kidney cancer cells.

Why is control of the cell cycle lost?

About 15% of human cancers are the result of viral infections which cause changes in the cells, affecting the control of the cell cycle and leading to cancer.

Most cancers result from mutations which occur in the DNA of a normal body cell as it reproduces:

- Several different mutations may work together to increase the likelihood of cancer, e.g. by interfering with the accurate replication of the DNA, decreasing the efficiency of the DNA repair mechanisms or increasing the likelihood of the chromosomes breaking during mitosis.
- A single mutation changing a proto-oncogene to an oncogene can cause cancer. Proto-oncogenes code for proteins which stimulate the cell cycle. Oncogenes produce uncontrolled amounts of stimulating proteins.
- Tumour-suppressor genes code for the production of chemicals which suppress the cell cycle, acting as a brake on cell division. If the gene mutates, the brake is removed and the cell just keeps on dividing.

What causes cancer?

Two main factors affect whether an individual will develop a form of cancer:

- Genetics: some people are more likely than others to experience a mutation which results in cancer, and some people are born with a mutation which gives them a very high risk of developing particular cancers.
- Environment: factors such as the tar in cigarette smoke, ionising radiation, the chemicals in alcoholic drinks and asbestos are carcinogenic – they increase the risk of mutations in vulnerable cells which can result in cancer.

If a genetically vulnerable person encounters carcinogenic environmental factors, their risk of developing cancer somewhere in their body increases even more.

Where else will I encounter these themes?

1.1   1.2   1.3   1.4   2.1   2.2   2.3   YOU ARE HERE   2

Let us start by considering the nature of the writing in this article.

There are many different ways of getting information across. This article assumes knowledge of cell division in its readers:

1. How well does a question and answer format work for imparting this information?
2. Are bullet points helpful or would continuous text make the explanations more effective?
3. Try a different approach – work with a partner and choose two alternative ways of getting this information across. Each take one method and work it through – then compare notes and decide which you prefer and why.
4. How would you explain cancer to people who are not studying AS/A level Biology and do not know what mitosis and the cell cycle are?

> Alternative ways of getting information across include formal continuous prose, a series of diagrams, flow diagrams, a more relaxed and friendly style of writing ... and any others you can think of! For a non-scientific audience you have to think about the words you use – and how pictures might help.

Now let us look at the biology behind these questions and answers. You already know about cell division and mutation, so you can answer these questions now. If you are going to continue your biology studies to A level, you may like to revisit these pages after you have learned more about communicable diseases and the immune system in **Book 2 Topic 6**.

5. At one level there are as many types of cancer as there are types of human cell – any type of cell can become cancerous, although some are more prone to cancer than others. At another level, every cancer is different because each one of us is different and it is our cells that mutate. Describe how the process of cell division can lead to mutations and therefore cancer.

> **Command word**
> If you are asked to describe something, you need to give a clear and concise account of it. Pull out all the important pieces of information as a starting point.

Activity

You are going to produce an article for a popular magazine explaining about one of the following forms of cancer. You must explain what happens to the cells of your body and why in each case the risk of developing cancer is increased. You can add as much extra information about detection, screening, prevention and even treatment as you like – but the key points to focus on are the biology of the cancer. Write about one of the following.

1 Breast cancer resulting from mutations in the BRCA1 and BRCA2 tumour suppressor genes.

2 Cervical cancer caused by the human papilloma virus.

● From an article by Ann Fullick.

1 The photograph below shows a cell in the metaphase stage of mitosis as seen using a light microscope.

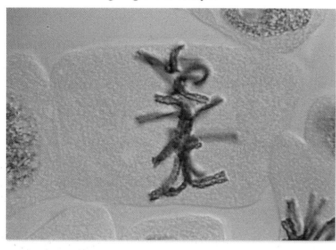

(a) Which of the following statements is correct?

A Metaphase occurs after anaphase and before telophase

B Metaphase occurs after prophase and before anaphase

C Metaphase occurs after telophase and before prophase

D Metaphase occurs after anaphase and before prophase

[1]

(b) Draw and label a diagram to show the appearance of a chromosome in metaphase. [3]

(c) Describe how the cell shown in the photograph would differ when it is in anaphase. [2]

[Total: 6]

2 (a) The cell cycle includes interphase and mitosis. Mitosis has four phases: prophase, metaphase, anaphase and telophase. The photograph below shows plant root cells at various stages of the cell cycle. *anaphase*

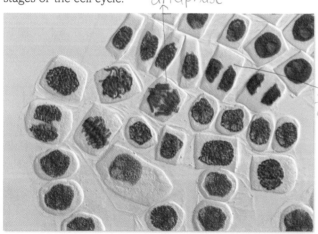

(i) Draw a line to indicate a cell in the photograph that is undergoing **anaphase** and label this line A. [1]

(ii) Draw a line to indicate a cell in the photograph that is undergoing **telophase** and label this line T. [1]

(iii) How many of the cells shown in the photograph are in **telophase**? [1]

(b) Give an account of the events that take place during prophase and metaphase of mitosis. [5]

[Total: 8]

3 The graphs below show changes in the DNA content of cells during the cell cycle in two different plants, **A** and **B**.

Plant A

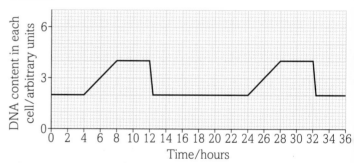

Plant B

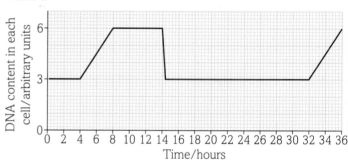

(a) Compare and contrast the cell cycle of **plant A** with the cell cycle of plant B. [3]

(b) The DNA content of the cells of **plant A** doubles between 4 and 8 hours. Give an explanation for this change in DNA content. [2]

(c) Describe the events that are occurring inside the cells of plant A between 11 and 13 hours. [2]

[Total: 7]

4 (a) The flow diagram below shows a method for preparing and staining cells in order to observe chromosomes.

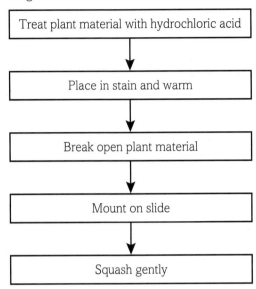

(i) Name a suitable part of a plant to use. Give a reason for your answer. [2]

(ii) Name a suitable stain for this method. [1]

(iii) Explain why it is necessary to squash the plant material at the end of the process. [1]

(b) The diagrams **A**, **B**, **C** and **D** below show four stages of mitosis.

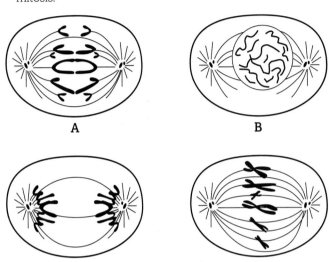

(i) Write the letters to show the correct sequence of the stages in mitosis. [1]

(ii) Name the stages shown in diagrams **A** and **C**. [2]

[Total: 7]

TOPIC 2
Cells and viruses

2.4 > Meiosis and sexual reproduction

Introduction

When we look at cell division we always talk about eukaryotic cells having two sets of chromosomes, one from each parent. Polysomy – having an extra copy of just one chromosome – is seen as an example of things going wrong. However, many plant species are actually polyploid – they have several complete sets of chromosomes! Increasingly, evidence suggests that polyploidy helps plants survive changing or adverse conditions. This may be why 40% of plant species survived the mass extinction event that wiped out the dinosaurs around 65 million years ago.

The spare genetic information in polyploidy may enable plants to overcome harmful mutations, and they can use the same gene to control different functions. In the Arctic Circle, many of the plants that survive best in the extreme conditions are polyploid – and it is the polyploids that are most successful in colonising bare areas left as glaciers retreat. There is even some evidence that polyploidy may have played a part in human evolution.

In this chapter you will be looking at meiotic cell division in eukaryotic cells. Meiosis is the process by which a reduction in the chromosome number is achieved in the formation of the sex cells or gametes. Meiosis only takes place in the sex organs.

You will learn the process of meiotic cell division and its importance in introducing genetic variation, key to survival in changing conditions, to natural selection and to evolution. You will also consider what happens in chromosome mutations, and in non-disjunction of the chromosomes resulting in either monosomy or polysomy.

You are going to look at gametogenesis – the formation of the sex cells – in both mammals and plants, and then go on to explore how those gametes join at fertilisation to form a new genetic individual. You will follow the mammalian zygote through the early stages of embryonic development until it implants in the uterus of the mother.

All the maths you need

- Carry out calculations using numbers in standard and ordinary form (*e.g. use of magnification*)
- Use scales for measuring (*e.g. measuring sizes of male and female gametes*)
- Find arithmetic means (*e.g. measuring sizes of male and female gametes*)
- Make order of magnitude calculations (*e.g. use and manipulate the magnification formula:*
  $$magnification = \frac{size\ of\ image}{size\ of\ real\ object})$$
- Use and manipulate equations, including changing the subject of an equation (*e.g. magnification*)

What will I study later?

- Mutations as a source of new variations
- Classification of plants involves the numbers and role of the cotyledons in the seed
- Post-transcriptional changes in mRNA can result in different products from the same gene (A level)
- Epigenetic modifications and their effect on totipotent stem cells in the embryo (A level)
- The role of random assortment and crossing over in meiosis in giving rise to new combinations of alleles in gametes (A level)
- How random fertilisation during sexual reproduction brings about genetic variation (A level)
- Autosomal and sex linkage of alleles (A level)

What have I studied before?

- The role of meiotic cell division in halving the chromosome number to form gametes
- The ultrastructure of eukaryotic cells including the nucleus, nuclear membrane, chromosomes, centrioles, etc.
- The way in which DNA replicates in the nucleus
- Gene mutations

What will I study in this chapter?

- The role of meiosis in the production of haploid gametes, including the stages of meiosis
- The replication and division of the genetic material in the main stages of meiosis
- The ways in which meiosis results in genetic variation through recombination of alleles, including independent assortment and crossing over
- Chromosome mutations such as translocations
- Non-disjunction of the chromosomes leading to conditions such as Down's syndrome and Turner's syndrome
- The development of the female and male gametes in mammals – oogenesis and spermatogenesis – and in plants with the formation of the pollen grain and the embryo sac
- Fertilisation in mammals followed by the early development of the embryo
- Fertilisation in plants including the roles of the tube nucleus, the pollen tube and enzymes, and the process of double fertilisation inside the embryo sac to form a triploid endosperm and the zygote

By the end of this section, you should be able to...

● describe the role of meiosis in the production of haploid gametes including the stages of meiosis

● explain how meiosis results in genetic variation through recombination of alleles, including independent assortment and crossing over

As you have seen, asexual reproduction can be very successful at producing new individuals, but leaves the population vulnerable to changes in the environment. The offspring are largely identical to their parents. There is very limited genetic variety, although spontaneous mutations do occur.

Relatively few organisms rely solely on asexual reproduction. Most have at the very least a back-up system of sexual reproduction, used when conditions are tough, to introduce the genetic variation that may enable the population or species to survive. In more complex organisms, particularly animals and flowering plants, sexual reproduction is the main way of producing fertile offspring.

Sexual reproduction is the production of a new individual resulting from the joining of two specialised cells known as gametes. The individuals that result from sexual reproduction are not genetically the same as either of their parents, but contain genetic information from both of them. Sexual reproduction relies on two gametes meeting and fusing. It is not always easy to find a mate, particularly if you are a solitary predator. It is also more expensive in terms of bodily resources because it usually involves special sexual organs. But the great advantage of sexual reproduction is that it increases genetic variation as a result of the fusing of gametes from two different individuals. In a changing environment, this gives a greater chance that one or more of the offspring will have a combination of genes that improves their chance of surviving and going on to reproduce.

fig A The genetic variation in offspring produced by sexual reproduction can be very easy to see.

What are gametes?

The nucleus of a cell contains the chromosomes. In most of the cells, the chromosomes occur in pairs. A cell containing two full sets of chromosomes is called **diploid (2n)** and the number of chromosomes in a diploid cell is characteristic for that species. However, if two diploid cells combined to form a new individual in sexual reproduction, the offspring would have four sets of chromosomes, losing the characteristic number for the species. Each new generation would become more heavily loaded with genetic material until eventually the cells would break down and fail to function. To avoid this, **haploid (n)** nuclei are formed with one set of chromosomes (half of the full chromosome number), usually within the specialised cells called gametes. Sexual reproduction occurs when two haploid nuclei fuse to form a new diploid cell called a **zygote** (see **fig B**), a process called **fertilisation**.

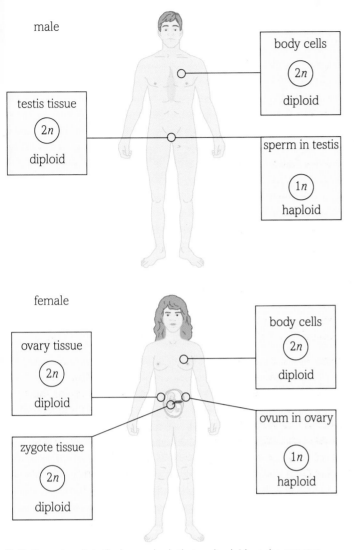

male

body cells

2n

diploid

testis tissue

2n

diploid

sperm in testis

1n

haploid

female

body cells

2n

diploid

ovary tissue

2n

diploid

zygote tissue

2n

diploid

ovum in ovary

1n

haploid

fig B The only cells in the human body that are haploid are the gametes.

Did you know?

Polyploidy

Although most eukaryotic organisms are diploid, a number have stable forms of **polyploidy**. Fish can be polyploid; for example salmon have four sets of chromosomes and some fish have as many as 400 chromosomes in total. Some reptiles are also polyploid. Polyploidy is very common in plants, particularly ferns and flowering plants. Potatoes are polyploid and so are the members of the genus *Dendranthema*. The haploid number of the genus is 9, but the polyploid numbers average 18 (2 sets of chromosomes) to 198 (22 sets of chromosomes).

Crop plants are often polyploid. Some wheat varieties are diploid, tetraploid and hexaploid, and the cabbage family is also hexaploid.

The formation of gametes

Gametes are formed in special sex organs. In simpler animals and plants the sex organs are often temporary, formed only when they are needed. In more complex animals the sex organs are usually more permanent structures that are sometimes called the **gonads**. In flowering plants the male sex organs are the **anthers** and the

female ones are the **ovaries**. The male gametes are formed in the **pollen**, produced in the anthers, and the female gametes are formed in the **ovules**, contained in the ovaries. In animals the male gonads are the **testes**, which produce the male gametes known as **spermatozoa** (**sperm**). The female gonads are the ovaries and they produce the female gametes known as **ova** or eggs. The male gametes are often much smaller than the female ones, but they are usually produced in much larger quantities. This can be summarised as:

- male: many, mini, motile
- female: few, fat, fixed.

Meiosis

In **Section 2.3.2** you saw that when cells divide by mitosis the number of chromosomes in both the daughter cells is the same as in the original parent cell. However, in the cell divisions that form gametes the chromosome number needs to be halved to give the necessary haploid nuclei. To bring about this reduction in the chromosome number, gametes are formed by a different process of cell division known as meiosis.

Meiosis is a reduction division and it occurs only in the sex organs. In animals the gametes are formed directly from meiosis. In flowering plants meiosis forms special male cells called **microspores** and female cells called **megaspores**, which then produce the gametes. Meiosis is of great biological significance – it is the basis of the variation that allows species to evolve.

What happens to the chromosomes?

In meiosis two nuclear divisions give rise to four haploid daughter cells, each with its own unique combination of genetic material (see **fig C**). The events of meiosis are continuous although we describe the stages as separate phases. As in mitosis, the contents of the cell, and in particular the DNA, are replicated while the cell is in interphase. Once the cell has all the materials it needs it can enter meiosis.

Many of the stages of meiosis are very similar to those of mitosis, with just a couple of crucial variations. The chromosomes replicate to form chromatids joined by a centromere as in mitosis. However, in meiosis the two chromosomes of each pair, known as **homologous pairs**, stay close together. At this stage **crossing over** or **recombination** takes place, introducing genetic variation as the chromatids break and recombine (see **fig E**). Just as in mitosis, the nuclear membrane and nucleolus break down and the centrioles pull apart to form the spindle. The centromeres do not split in the first division of meiosis, so pairs of chromatids move to the opposite ends of the cell. The cell then immediately goes into a second division without any further replication of the chromosomes. This division is just like mitosis. The centromeres divide and chromatids move to opposite poles of the cell. Finally the nuclear membranes re-form as the chromosomes decondense and become invisible again. Cytokinesis takes place giving four haploid daughter cells, each with half the chromosome number of the original parent cell. These daughter cells later develop into gametes.

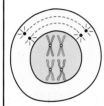
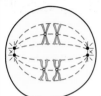
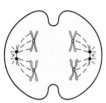

Prophase 1 – each chromosome appears in the condensed form with two chromatids. Homologous pairs of chromosomes associate with each other. **Crossing over** occurs.

Metaphase 1 – the spindle forms and the pairs of chromosomes line up on the metaphase plate.

Anaphase 1 – the centromeres do not divide. One chromosome (pair of chromatids) from each homologous pair moves to each end of the cell. As a result the chromosome number in each cell is half that of the original.

Telophase 1 – the nuclear membrane re-forms and the cells begin to divide. In some cells this continues to full cytokinesis and there may be a period of brief or prolonged interphase. During this interphase there is *no further replication* of the DNA.

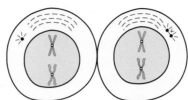
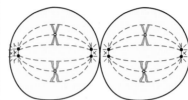

Prophase 2 – new spindles are formed.

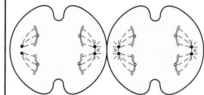

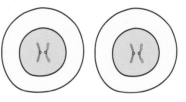

Metaphase 2 – the chromosomes, still made up of pairs of chromatids, line up on the metaphase plate.

Anaphase 2 – the centromeres now divide and the chromatids move to the opposite ends of the cell.

Telophase 2 – nuclear envelopes re-form, the chromosomes return to their interphase state and cytokinesis occurs, giving four daughter cells each with half the chromosome number of the original diploid cell.

fig C The main steps in the process of meiosis, which results in the formation of haploid gametes. This is a simplified version of meiosis shown in a cell with only two pairs of chromosomes.

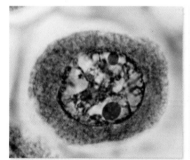
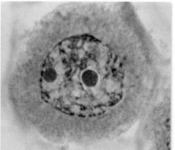
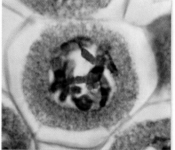
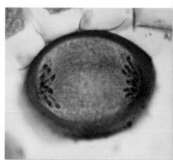

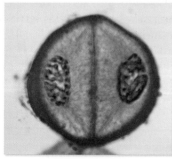

fig D The stages of meiosis are not easy to see in cells, but these images from the testis of a locust and anther of a plant show you many of them.

The importance of meiosis

Meiosis reduces the chromosome number in gametes from diploid to haploid, so that sexual reproduction is possible without each generation carrying an increasing burden of genetic material. It is also the main way in which genetic variation is introduced to a species. This variation is introduced in two main ways:

- **Independent assortment (random assortment)**: the chromosomes that came from the individual's two parents are distributed into the gametes and so into their offspring completely at random. For example, each gamete you produce receives 23 chromosomes. In each new gamete any number from none to all 23 could come from either your maternal or your paternal chromosomes. It has been calculated that there are more than eight million potential genetic combinations within the sperm or the egg. This alone guarantees great variety in the gametes.

- **Crossing over (recombination)**: this process takes place when large, multi-enzyme complexes 'cut and join' bits of the maternal and paternal chromatids together. The points where the chromatids break are called **chiasmata**. These are important in two ways. First, the exchange of genetic material leads to added genetic variation in its own right. Second, errors in the process lead to mutation (see **Sections 1.3.6** and **2.4.2**) and this is a further way of introducing new combinations into the genetic make-up of a species.

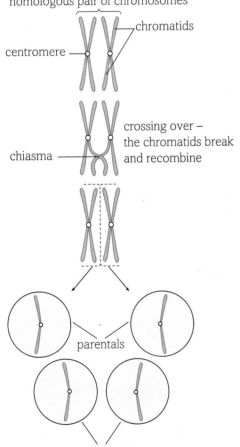

homologous pair of chromosomes

chromatids

centromere

crossing over – the chromatids break and recombine

chiasma

parentals

recombinants

fig E Chromosomes crossing over in meiosis – this process introduces more variation into the gametes.

Questions

1 Give examples of conditions when sexual reproduction would be more advantageous in the production of offspring than asexual reproduction, and vice versa. Explain your choices.

2 Make a table to summarise the main stages in the process of meiosis.

3 Explain how meiosis leads to variation between offspring.

Key definitions

Diploid (2n) signifies a cell with a nucleus containing two full sets of chromosomes.

Haploid (n) signifies a cell with a nucleus containing one complete set of chromosomes.

A **zygote** is the cell formed when two haploid gametes fuse at fertilisation.

Fertilisation is the fusing of the haploid nuclei from two gametes to form a diploid zygote in sexual reproduction.

Polypoidy is when a cell or an organism has more than two sets of chromosomes.

Gonads are the sex organs in animals.

Anthers are the male sex organs in plants.

Ovaries are the female sex organs in both animals and plants.

Pollen, produced in the anthers, contains the haploid male gametes in plants.

Ovules, formed in the ovaries, contain the haploid female gametes in plants.

Testes are the male sex organs in animals.

Spermatozoa (sperm) are the haploid male gametes in animals.

Ova are the haploid female gametes in animals.

Microspores are the result of meiosis in plants that produce the male gametes, pollen.

Megaspores are the result of meiosis in plants that produce the female gametes, ovules.

Homologous chromosomes describe a set of one maternal chromosome and one paternal chromosome that pair up during meiotic cell division.

Crossing over (recombination) is the process by which large multi-enzyme complexes cut and rejoin parts of the maternal and paternal chromatids at the end of prophase 1.

Independent assortment (random assortment) is the process by which the chromosomes derived from the male and female parent are distributed into the gametes at random.

Chiasmata are the points where the chromatids break during recombination.

Mutations

By the end of this section, you should be able to...

● describe chromosome mutations, illustrated by translocations

● explain how non-disjunction can lead to polysomy and monosomy

In **Section 1.3.6** you learned about gene mutations and the way these can affect the proteins produced by the cells in protein synthesis. However, these are not the only forms of mutations that can affect the genetic material during the process of replication and meiosis.

Chromosome mutations

Sometimes during the process of meiosis, parts of the chromosomes break off and become reattached in the wrong place. These are called chromosome mutations. They are changes on a larger scale than the gene mutations you looked at in **Section 1.3.6**. One of the most common forms of a chromosome mutation is **translocation**. This takes place when a piece from one pair of homologous chromosomes breaks off and reattaches to one of a completely different pair of chromosomes.

Some translocations are balanced – a piece is effectively swapped between two different chromosomes (see **fig A**). People who have balanced translocations are often healthy. However, some translocations are unbalanced – one chromosome loses a piece and another chromosome gains it. These mutations can cause big changes to the phenotype of the individual. They may even be incompatible with life.

If a point or gene mutation is like changing a letter in a word (see **Section 1.3.6**), a chromosome mutation is like changing or removing a word, or completely rearranging the words in the sentence. If we are lucky they will still make sense, but they will probably not mean the same thing as the original sentence.

Translocations can have severe effects. For example, translocations between chromosome 8 and chromosome 21 can result in a type of blood cancer known as core binding factor acute myeloid leukaemia, whilst translocations between chromosome 8 and chromosome 14 can cause Burkitt's lymphoma, a cancer of white blood cells mainly seen in children and young adults.

fig A In a balanced chromosome translocation, part of one chromosome is swapped with a section from a completely different pair of chromosomes.

Non-disjunction of the chromosomes

Some mutations affect not single genes or parts of chromosomes, but whole chromosomes. When the cells in the ovaries and testes undergo meiotic division to form the ova and sperm, the chromosome number in the cells is halved. In humans, each gamete should contain 23 chromosomes, including one **sex chromosome**, to pass on to the next generation. However, sometimes an error called **non-disjunction** occurs. During the reduction division of meiosis, the members of one of the homologous pairs of chromosomes fail to separate during anaphase 2. As a result, one of the gametes has two copies of that chromosome, and another has no copies. If one of these abnormal gametes joins with a normal gamete and is fertilised, the individual who results will either have **monosomy** with only one member of the homologous pair present from the normal gamete, or **polysomy** with three or more rather than two chromosomes of a particular type. The situation where a cell either lacks a whole chromosome or has more than two of a chromosome is called **aneuploidy**. Most examples of aneuploidy are fatal and affected fetuses do not develop to term. For example, trisomy of chromosome 16 sometimes occurs, but all of the embryos abort spontaneously in the first three months of pregnancy. Trisomy of chromosome 18 causes Edward's syndrome. Most affected babies are stillborn, and those that are born alive almost always die within a year. However, there are some examples of aneuploidy where those affected can and do survive, particularly when the sex chromosomes are affected rather than the autosomes.

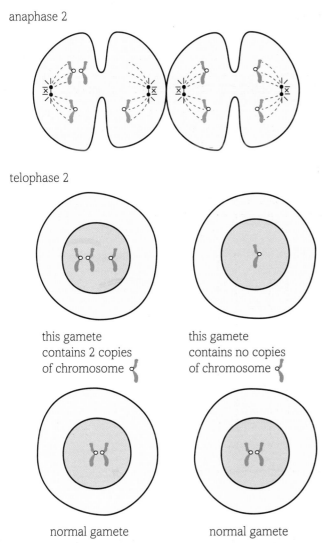

fig B Non-disjunction during meiosis results in gametes with an extra chromosome or a missing chromosome.

Example 1: Down's syndrome (polysomy)

If there is non-disjunction of chromosome 21 in an ovum or sperm, one of the gametes will contain two copies of the chromosome. After fertilisation with a normal gamete, the resulting zygote will have polysomy, with three copies of chromosome 21. Babies born with trisomy of chromosome 21 have Down's syndrome. The extra chromosome affects both mental and physical development. Children born with Down's syndrome will often have problems including heart abnormalities, learning difficulties that can be quite severe, lack of muscle tone and visual problems.

fig C The physical and mental characteristics of Down's syndrome result from one extra copy of chromosome 21.

Did you know?

Factors affecting mutation rates

Mutations are relatively rare events. However, certain factors increase the frequency with which they occur. Ionising radiation is known to increase the **mutation rate**. Certain chemicals (called **mutagens**) also increase the rate of mutation. The rate of non-disjunction increases with the age of the parents – both monosomy and polysomy are more common in children born to older mothers and fathers. The ova are particularly vulnerable. All of the future ova of a woman are suspended in prophase 1 of meiosis while the woman herself is still an embryo. Meiosis is completed only after ovulation and fertilisation by a sperm (see **Section 2.4.1**). The ova are exposed to all the chemical and radiation hazards of modern life and as a result the rate of non-disjunction increases as the woman gets older.

The effect is clearly seen in the increased incidence in Down's syndrome pregnancies as women get older. The risk of having a baby with Down's syndrome for a 20-year-old woman is 1 in 1667. As she approaches 50 – about the limit of natural child-bearing – the risk has increased. Figures quoted range from 1 in 5 to 1 in 18 births. The same pattern is seen for other conditions involving non-disjunction of the chromosomes.

Example 2: Turner's syndrome (monosomy)

Apart from children affected by Down's syndrome, few fetuses with aneuploidy of the autosomes survive even until birth. However, the absence of a sex chromosome or the presence of extra sex chromosomes is less unusual and less life-threatening, although in most cases it does cause fertility problems. The presence or absence of a Y chromosome determines the route for sexual development in human embryos. Any embryo with at least one Y chromosome will develop male characteristics, whilst any embryo lacking a Y chromosome will develop female characteristics. When there is non-disjunction of the male sex chromosomes, an egg may be fertilised by a sperm that has no sex chromosomes. The resulting embryo will have monosomy – just one X chromosome from the ovum. They will have the genotype XO that results in Turner's syndrome. The affected person is apparently female, but she is infertile and will not undergo puberty without being given extra sex hormones. However, non-disjunction of the male chromosomes will also give sperm carrying both an X and a Y chromosome. If one of these sperm fertilises a normal egg the resulting embryo will be XXY. This is called Kleinfelter's syndrome and affected individuals have small testes and produce little testosterone. They have little facial and body hair, may develop breast tissue, have less muscle development than usual and may be infertile. Around 1 in every 600 live male births is affected by Kleinfelter's syndrome, and the individuals may be indistinguishable from their peers or relatively severely affected.

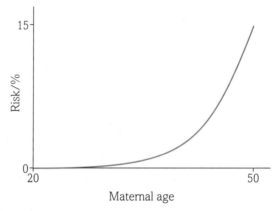

fig D This graph shows how the risk of having a child with Down's syndrome increases with the age of the mother.

Questions

1   Explain the difference between crossing over and translocation in meiosis and the effects they have on the organism involved.

2   What is the difference between polysomy and monosomy and how do they occur?

Key definitions

A **translocation** (noun) is a mutation in which part of one chromosome breaks off and rejoins to another completely different chromosome. It may be balanced, if part of two chromosomes effectively swap, or it may be unbalanced if a piece simply breaks off one chromosome and joins another.

Sex chromosomes are the chromosomes that carry the information that determines the sex of the individual. Human females have two X chromosomes (XX) and males have an X and a Y chromosome (XY).

Non-disjunction is the process that occurs when members of a pair of chromosomes fail to separate during the reduction division of meiosis, resulting in one gamete with two copies of a chromosome and one gamete with no copies of that chromosome.

Monosomy is when only one member of a pair of chromosomes is present in a cell.

Polysomy is when a cell contains three or more rather than two chromosomes of a particular type.

Aneuploidy is when a cell contains too few or too many chromosomes.

The **mutation rate** is the rate at which mutations naturally occur.

Mutagens are chemicals known to increase the rate of mutation.

By the end of this section, you should be able to...

● explain the process of oogenesis and spermatogenesis in mammals

● explain how a pollen grain forms in the anther and the embryo sac forms in the ovule in plants

The gametes that make sexual reproduction possible are formed in a process called **gametogenesis**. Meiosis is just one stage in gamete formation, which produces different male and female sex cells. You are going to consider the way in which sperm and ova are made in the sex organs of mammals, using humans as an example, and also how gametes are formed in flowering plants.

Gametogenesis in mammals

Many millions of sperm are released every time a male mammal ejaculates. The eggs in a sexually mature female are usually numbered in thousands and will eventually run out. Special cells (the **primordial germ cells**) in the gonads divide, grow, divide again and then differentiate into the gametes.

Both mitosis and meiosis play a role in gametogenesis. Mitosis provides the precursor cells. Meiosis brings about the reduction divisions that result in gametes. In human males, the process of gametogenesis continues constantly from puberty. In females, the mitotic divisions take place before birth. The meiotic divisions take place in a few oocytes in each monthly cycle from puberty to menopause and are only completed if the oocyte is fertilised.

Spermatogenesis

Spermatogenesis is the formation of spermatozoa. Each primordial germ cell in the testes results in large numbers of spermatozoa (see **fig A**). There are enormous numbers of primordial germ cells in the testes producing millions of spermatozoa on a regular basis:

- The diploid primordial germ cell divides several times by mitosis to form diploid spermatogonia.
- The spermatogonia then grow without further division until they are big enough to be called primary spermatocytes.
- The spermatocytes undergo meiosis. The first meiotic division results in two haploid cells called secondary spermatocytes.
- The second meiotic division results in four haploid cells called spermatids.
- The spermatids then differentiate in the tubules of the testes to form spermatozoa, the active gametes capable of fertilising an ovum.

Oogenesis

Oogenesis is the formation of ova. Each primordial germ cell in an ovary results in only one ovum (see **fig A**). As a result the number of female ova is always substantially smaller than the number of spermatozoa. Ova contain a much higher proportion of material than sperm so there is a much greater investment of resources in each one. It would not make sense biologically to waste resources by producing too many of them:

- The diploid primordial germ cell divides several times by mitosis to form diploid oogonia. Most of the oogonia do not develop further. They simply degenerate. Only one continues to grow and substantial amounts of storage material go into the cell making it very large compared with the spermatocytes. At this stage the large cell is known as a primary oocyte.
- The oocyte undergoes meiosis. The first meiotic division results in two cells of very unequal size. The larger cell is the secondary oocyte. The other, much smaller, cell sticks to the oocyte and is called the first polar body. At this stage the oocytes do not divide further until after ovulation. What we call ova in the ovary are really secondary oocytes.

• The second meiotic division is only completed after fertilisation occurs. The secondary oocyte divides to form the haploid ovum and another polar body, whilst the first polar body divides to form two more polar bodies. The polar bodies do not seem to have any function except to receive the chromosomes in the meiotic divisions. They degenerate and die as the ovum develops.

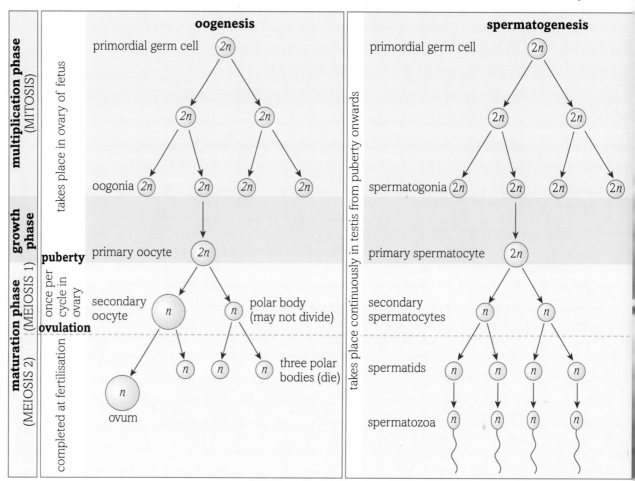

fig A The formation of the sperm in the testis and the ova in the ovary.

Characteristics of the gametes

Spermatozoa: many, mini, motile

The male gametes or spermatozoa of most mammalian species, including humans, are around 50 μm long. They have several tasks to fulfil. They must remain in suspension in the semen so they can be transported through the female reproductive tract, and they must be able to penetrate the protective barrier around the ovum and deliver the male haploid genome safely inside. The close relationship between the human spermatozoan's structure and its functions is shown in **fig B**. Millions upon millions of these motile gametes are produced in the lifetime of a human male. The average size of a UK family is about 1.7 children, with only one spermatozoan needed to fertilise each ovum, so this gives an idea of the scale of biological wastage.

Ova: few, fat, fixed

Although spermatozoa of most animals are very similar in size, the same cannot be said for ova. These vary tremendously in both their diameter and their mass. The human ovum is about 0.1 mm across, whilst the ovum contained in an ostrich egg is around 6 mm in diameter. Eggs do not move on their own, so they do not need contractile proteins, but they usually contain food for the developing embryo. The main difference between eggs of various species is the quantity of stored food they contain. In birds and reptiles a lot of development takes place before the animal hatches, so the egg contains a large food store. In mammals, once the developing fetus has implanted in the uterus it is supplied with nutrients from the blood supply of the mother and so large food stores in the egg are unnecessary.

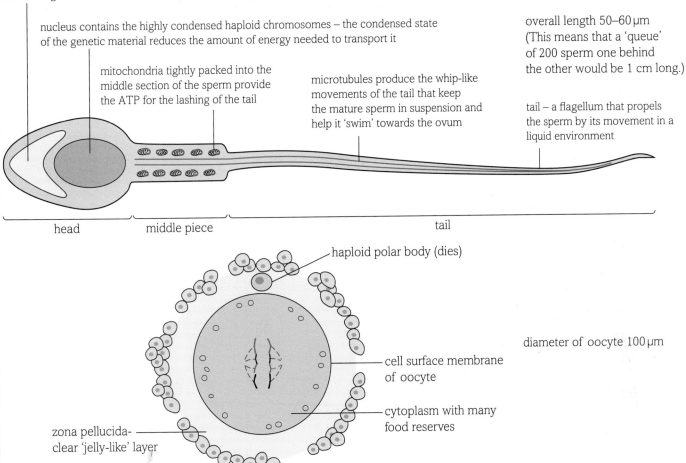

acrosome – membrane-bound storage site for enzymes that digest the layers surrounding the ovum and allow the sperm's head to penetrate

nucleus contains the highly condensed haploid chromosomes – the condensed state of the genetic material reduces the amount of energy needed to transport it

overall length 50–60 μm (This means that a 'queue' of 200 sperm one behind the other would be 1 cm long.)

mitochondria tightly packed into the middle section of the sperm provide the ATP for the lashing of the tail

microtubules produce the whip-like movements of the tail that keep the mature sperm in suspension and help it 'swim' towards the ovum

tail – a flagellum that propels the sperm by its movement in a liquid environment

head

middle piece

tail

haploid polar body (dies)

diameter of oocyte 100 μm

cell surface membrane of oocyte

cytoplasm with many food reserves

zona pellucida- clear 'jelly-like' layer

fig B Human gametes – the sperm and ova show clear specialisations that fit them for their function. They are not drawn to scale.

Gametogenesis in plants

The formation of gametes in flowering plants is more complex because plants have two phases to their life cycles. The **sporophyte generation** is diploid and produces spores by meiosis. The **gametophyte generation** that results is haploid and gives rise to the gametes by mitosis. In plants such as mosses and ferns, these two phases exist as separate plants. In flowering plants, the two phases have been combined into one plant. The main body of the plant that we see is the diploid **sporophyte**. The haploid gametophytes are reduced to parts of the contents of the anther and the ovary. They are produced by meiosis from spore mother cells.

The formation of pollen

The anthers of flowering plants are analogous to the testes of animals. Meiosis occurs here, resulting in vast numbers of the pollen grains that carry the male gametes. Each anther contains four **pollen sacs** where the pollen grains develop. In each pollen sac there are large numbers of microspore mother cells. These are diploid. They divide by meiosis (see **fig C**) to form haploid microspores, which are the gametophyte generation. The gametes themselves are formed from the microspores by mitosis. They contain two haploid nuclei, the **tube nucleus** and the **generative nucleus**. The tube nucleus has the function of producing a **pollen tube** that penetrates through stigma, style and ovary and into the ovule. The generative nucleus then fuses with the nucleus of the ovule to form a new individual.

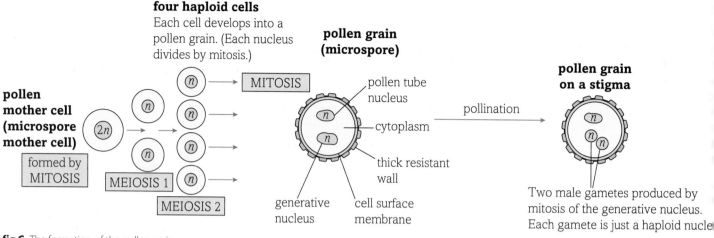

four haploid cells
Each cell develops into a pollen grain. (Each nucleus divides by mitosis.)

pollen grain (microspore)

pollen mother cell (microspore mother cell)

formed by MITOSIS

MEIOSIS 1

MEIOSIS 2

MITOSIS

pollen tube nucleus

cytoplasm

thick resistant wall

generative nucleus

cell surface membrane

pollination

pollen grain on a stigma

Two male gametes produced by mitosis of the generative nucleus. Each gamete is just a haploid nucleus.

fig C The formation of the pollen grain.

Did you know?

The pollen record

The surface patterns of pollen grains are unique and specific to the species. They are extremely tough and resistant to decay and can remain in the soil for thousands of years. Palaeobotanists can tell what plants were growing thousands of years ago and how abundant they were by analysing the pollen they find in archaeological digs.

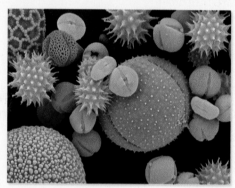

fig D These amazing looking pollen grains are from plants including sunflowers, morning glory, hollyhocks, lilies, primroses and castor oil plants.

The formation of egg cells

The ovary of the plant is analogous to the animal ovary. Meiosis results in the formation of a relatively small number of ova contained within ovules inside the ovary. Some plants – an example is the nectarine – produce only one ovule (egg chamber), whilst others such as peas produce several. The ovule is attached to the wall of the ovary by a pad of special tissue called the **placenta**. A complex structure of integuments (coverings) forms around tissue known as the nucellus. In the centre the embryo sac forms the gametophyte generation (see **fig E**).

Diploid megaspore mother cells divide by meiosis to give rise to four haploid megaspores, three of which degenerate leaving one to continue to develop. The megaspore undergoes three mitotic divisions that result in an embryo sac containing an egg cell, two polar nuclei and various other small cells (synergids and antipodal cells).

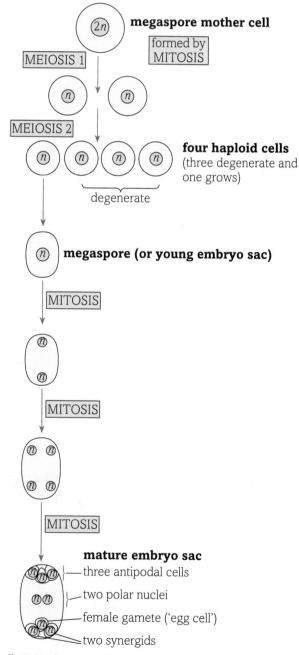

megaspore mother cell

formed by MITOSIS

MEIOSIS 1

MEIOSIS 2

four haploid cells (three degenerate and one grows)

degenerate

megaspore (or young embryo sac)

MITOSIS

MITOSIS

MITOSIS

mature embryo sac
— three antipodal cells
— two polar nuclei
— female gamete ('egg cell')
— two synergids

fig E The formation of the egg cell.